The Hidden Life of Deer

Also by Elizabeth Marshall Thomas

The Harmless People

Warrior Herdsmen

Reindeer Moon

The Animal Life

The Hidden Life of Dogs

The Tribe of Tiger: Cats and Their Culture

The Social Lives of Dogs: The Grace of Canine Company

Certain Poor Shepherds

The Old Way: A Story of the First People

The Hidden Life of Deer

Lessons from the Natural World

Elizabeth Marshall Thomas

HARPER
An Imprint of HarperCollins*Publishers*
www.harpercollins.com

HarperCollins books may be purchased for educational, business, or sales promotional use. For information, please write: Special Markets Department, HarperCollins Publishers, 10 East 53rd Street, New York, NY 10022.

FIRST EDITION

Designed by Leah Carlson-Stanisic

Library of Congress Cataloging-in-Publication Data

Thomas, Elizabeth Marshall, 1931–
The hidden life of deer : lessons from the natural world / Elizabeth Marshall Thomas. — 1st ed.
p. cm.
ISBN 978-0-06-179210-6
1. Deer—Behavior. 2. Thomas, Elizabeth Marshall, 1931– I. Title.

QL737.U55T4826 2009
599.65'15—dc22

2009002777

---

09 10 11 12 13 OV/RRD 10 9 8 7 6 5 4 3 2 1

This book is dedicated to my granddaughters, Zoë, Ariel, and Margaret; to my grandsons, David and Jasper; and to my great-grandson, Jacoby. I may have other great-grandchildren in time, and this book is dedicated to them too, but I cannot name them here because they have yet to be born. May the natural world in all its present wonder be there for them and for all children, grandchildren, and great-grandchildren.

Camping Alone

My friend Antler spends weeks alone in the wilderness every fall.
I have never spent any time camping alone, maybe two or three nights—
once when I got lost
and after wandering around the woods for two hours in the dark
I just lay down and slept in the leaves.
Antler talks about having to get used to walking on two legs again
when he returns.
He says that every year he leaves a little more of himself in the woods,
and that someday there will be more of him out there than here—
I think it may already have happened.
Someday, maybe—I'll go to some lonely spot and pitch my tent
and spend my days doing what one does when alone in the woods
and sleep night after night under the ten thousand stars.
But not in winter. Another guy disappeared in the Adirondacks last week—
it happens every year or two.
But not before he froze to death.
Solitary heart attack while temperature, snow, and night were falling.

Howard Nelson

Contents

Preface
A Note to Readers

The notion to write this book came from feeding corn to deer in the winter in south-central New Hampshire. I didn't know much about them except that they seemed to like corn. I wanted to know more. But they all looked alike and they wouldn't stand still, so it took a while to fathom their behavior and really see them. I have a friend, Katy Payne, who studies elephants. It was she who discovered that elephants make infrasound, at a time when such a thing seemed impossible and no other land mammals were known to do so. Katy once told me that her advice to students who are eager to join her research team is to start by studying deer. Deer are within a mile of almost everybody, and from them one can gain an understanding of what it's like to try to learn from wildlife. I thought of that as I watched my deer. For a while I studied wild elephants with Katy and was awed by her ability to recognize individuals. I could do that too, if not as well, but while I was trying to identify deer, it certainly seemed that elephants were easier.

I wished that Katy could see my deer. If anyone could sort them out, she could. I started to write her a letter to describe what they were doing. Soon the letter was many pages long and I saw it was a book, this book. So I continued with what seemed to me like research, and began to realize something that I certainly should have known already, that as members of the enormous deer family with its forty-odd species, whitetails were not unlike the other deer all over the world—mule deer of the West and Southwest, red deer or elk of the Holarctic, also India, Sri Lanka, and Burma, and reindeer or caribou of the far north, to say nothing of moose, the biggest of the deer, and all the different kinds of little deer in thickets and marshes from China to Argentina. I entered a realm full of cooperation, hierarchy, and a clear set of rules. Some of these rules are specific to whitetails, but many are fundamental to all groups, not only to the deer but also to all other creatures, which would include, of course, ourselves.

When I was a child, my father told me the chemical formulas for hemoglobin and chlorophyll, the substance in plants that makes them green and enables them to take in carbon dioxide and release our all-important oxygen. I remember the formulas to this day because they were almost identical—$C_{55}H_{72}FeN_4O_6$ was hemoglobin, he told me, and $C_{55}H_{72}MgN_4O_6$ was chlorophyll. Both had the same amounts of carbon, hydrogen, nitrogen, and oxygen, and only one difference: right in the middle where hemoglobin has iron, chlorophyll has magnesium.

Oh wow! Young as I was, that seemed important. That a plant and a person shared something so basic seemed awesome. That plants made the oxygen in every breath we drew also seemed awesome. It gave a sense of the oneness of things. It also gave a sense of what Nature had been able to do with that formula—making so many kinds of animals and plants, all of them different. At least to me, it gave a sense of our place on the planet. I saw that animals were important. I saw that plants were even more important. I was also to learn that compared to many of the other species, we weren't important at all except for the damage we do. We do not rule the natural world, despite our conspicuous position in it. On the contrary, it is our lifeline, and we do well to try to understand its rules.

It is also full of wonder, and it's right outside your door, perhaps even inside your house, even if you live in the city. You may be thinking that I'm just talking about mice and rats. And yes, I would include mice and rats. In fact, we're related to them via our common ancestor in the Cretaceous. They live by the rules of the natural world no matter how much we discourage them, and they do so very successfully. Our species has also been very successful. The ancestral stem must have been strong.

If we can forget our preconceptions and start fresh, observing any resident of the natural world as carefully as we can, trying to figure out what it's doing and why, we will see things we otherwise could not imagine. We can enter a world as different from ours as it's possible to be, the world to which we once belonged, a world we normally don't notice but which is all around us. We can't readily observe the burrowing insects in the soil, for instance, essential as they are to our well-being, and watching a plant until it does something perceptible can take a really long time, but we can easily observe many kinds of animals, especially birds and mammals who, because they are in ways so like ourselves, have much to show us.

Chapter One

The Year without Acorns

It began with a bird feeder by the kitchen door. The chickadees chose only the sunflower hearts and threw all other seeds to the ground where three gray squirrels and a red squirrel ate them. One day the seeds were discovered by a passing flock of nine or ten wild turkeys. My husband and I were thrilled to see wild turkeys near the house. I put out a little corn for them. Soon, a flock of twenty-eight turkeys came for the corn. What should I have done? Refused to feed the others? I fed the others. By the end of that winter I was feeding fifty-three turkeys.

We rarely saw the turkeys in the summer. They were finding food in the woods and also in our field, where long grass hid them. But they came to our house again in the fall, just the small, original flock at first, then other flocks, every morning just before dawn. Their calls would wake me, and I would bring out a pail of corn. My presence would scare them, and they would fly away. This was distressing. It takes energy to launch a bird the size of a turkey, and more energy if she must leap straight up into the air without first running a little way to gain momentum. The corn I offered was wasted if the turkeys had to spend their calories in unnecessary flying. I began putting the corn out at night, so it would be ready for them in the morning.

One dark night after the first snowfall, I went out wearing a white bathrobe. I had no reason to make noise and therefore went quietly, and to my surprise I found myself right next to three deer. Because I was in white against the white snow, they didn't pay much attention at first, and then moved off without panic, so I distributed the corn I was carrying and went back for more. After that the deer came often. Never again did they let me near them, but I watched them from the window after the moonlight returned.

The winter of 2007–2008 was hard on wildlife in our area. The acorn crop, which fattens many animals in the fall and feeds them in the winter, was almost nonexistent. The oak trees bore nothing but a few miserable acorns that were literally smaller than peas, cup and all. A knowledgeable friend told me something very interesting, which is that nut trees do this from time to time in order to cut down on their predators. If the trees always produced a standard crop of nuts, the animals who eat them would increase in number until they reached the carrying capacity of the environment, and were eating every nut that fell. The trees would have no chance to reproduce. To handle the problem, the trees hold back and let the animals starve. The oaks held back in 2007, creating a dangerous hardship for turkeys, deer, bears, and many others.

We are told not to feed any kind of wildlife, especially not deer. Why not? The naysayers have many reasons, the generic one being that the population of any wild species is formatted to its natural food supply, and to interfere with this is to enable more animals to live than the environment can sustain, causing the entire population to suffer from malnutrition. In other words, wild animals should starve naturally. And they do. However, the rationale is true of all living creatures, including small birds. Yet maintaining a bird feeder is laudable, and is practiced by many of those who tell you not to feed wild animals. Some species are consistent. Ours is not.

Thus, since very few people knowingly feed invertebrates or snakes or fish or frogs, the true meaning of the don't-feed rule is that you may feed little birds but not other birds and not mammals. Above all, you must not feed animals whose species are conspicuously successful. My mother found this out by feeding city pigeons and squirrels. She lived in Cambridge on a quiet street and put bread crumbs for birds out her kitchen window on the roof of her shed. At first, her crumbs attracted mostly English sparrows, but soon enough, pigeons and squirrels also found the food. They were just as hungry as the sparrows, and some of them, especially the squirrels, would look hopefully at my mother through the window. Regarding them as individuals in need, rather than as the collective symbol of a perceived ecological problem, she put out more food.

Her saintly neighbors were also her friends and made no complaint (at least not to her), even though, after years of bird and squirrel feeding, the roofs of their homes were sometimes loaded with pigeons, even though the sky could almost be darkened with pigeons, even though squirrels rushed down from every fence and tree when my mother scattered crumbs from her window. Thus I did not understand the depth of animosity toward people like my mother until the day she worried that she wasn't feeding them enough. She phoned the Audubon Society (to which she gave generous contributions) and asked how much food a city pigeon should eat. The response shocked and upset her. The rottweiler who took her call berated her so fiercely that my poised and worldly mother gasped and became a little shaky. Before this, I also had contributed to the Audubon Society, but their attack on my mother distressed me. After that, their appeals for funds evoked the needless pain they had caused a kind, caring woman. I threw away the appeals unopened, and in time, the Society stopped entreating me.

Why are there so many pigeons? The birds we call city pigeons are also called rock doves because they once nested only on cliffs. Originally these birds were uncommon, as they were brought to the New World by early immigrants from Europe, and not many nesting sites were available. However, rock doves can eat almost anything, and with the advent of cities, they found they didn't need cliffs after all, because they could just as well nest on the ledges of tall buildings. So, with plenty of edible trash and places to raise their young, their populations grew. Today, of course, they number in the billions, as did their close relatives, the now-extinct passenger pigeons, before we exterminated them. Do we begrudge passenger pigeons their former large numbers? Far from it. How sublime, we think, to see a migration of passenger pigeons filling the sky from horizon to horizon! If only they had not gone extinct!

We have no such admiration for the birds we once called rock doves. They're here in real life, not just in Audubon's drawings, so to us they are pests. We are offended by animals who are too plentiful, and we rename them pejoratively, usually for rats, who also are considered to be too plentiful. Hence pigeons are called flying rats.

The rules about animal feeding may vary, but my mother was steady and true. Although saddened by her Audubon attacker, she continued to feed her flocks to the best of her ability. She lived in that house for more than seventy years, and hundreds of individual birds and squirrels owed their lives to her compassion. More she couldn't do. At the age of ninety-eight she came to live with me in New Hampshire. As we arranged her move, the prospect of leaving behind the animals who depended on her was troubling. We considered live-trapping as many of her squirrels as possible and bringing them with us but gave up the plan in the end, partly because the newcomers would only displace the local population—that, or succumb themselves—and partly because one of her saintly neighbors offered to feed squirrels for her. At any rate, we had deer and turkeys, and it gave my mother much joy to see them.

As for interfering with nature, we have done plenty of that during our sojourn on this planet, to the detriment of nearly everything else, so like my mother, I see no reason, at this point in our history, not to offer the occasional helping hand to an animal in need, especially when the oaks decide to further their reproductive interests as they did in 2007. Near my house, I feed the oak trees too.

I came to New Hampshire early in life, and live in a house built in 1935 by my father on land in Peterborough that we once farmed. I've been watching the wildlife ever since. Back in the 1930s, farming was so pervasive that there were few if any deer, but their population increased, and today they graze in our fields in spring and summer, and sleep there too, in beds they make in the grass on top of the hill. When they hear us come home in the car at night, they stand up, and we see their eyes shining.

Although even today in southern New Hampshire it's rare to see deer in the woods—which is something to think about, considering that they are easily the most abundant of the large North American mammals—you can see the trails they make and follow, or where they rub their foreheads on trees or make scrapes or take shelter, or eat twigs or bark in winter, or make yards in deep snow. I am astonished by their intelligence, and their intricate knowledge of their world. Sometimes a few deer will leave the woods and start across the field, only to notice something unpleasant and run back to the cover of the trees. Once hidden, they walk onward unafraid. This doesn't seem like much until you realize that if you were standing with them in the woods, you could see everything around you and also everything in the field. You would feel fairly exposed, although to a distant observer, you would not be. How do the deer know they are hidden?

Those of us who spend most of our time in automobiles and buildings won't find an easy answer, and to appreciate what the deer were doing we would need to enter a woods from a field and try to figure out at which point we could no longer be seen. Only by knowing the woods very exactly would the vanishing point be apparent. Fleeing deer reenter the woods in all sorts of places, slowing down or stopping right after they know they have disappeared—except to an observer with high-powered binoculars. To me, this says that they have studied their world so carefully that they know all its properties. I'd say they know their territories better than many people know their houses.

One day in the woods a doe stood up ahead of me and slipped over the ridge I was climbing. I went to the place where she had been resting, an inconspicuous dent in the ground below the branches of a hemlock tree. The place was still warm from her body, I found when I lay down there to see what she had been seeing, and I marveled at her flawless choice—she herself had been invisible under the branches, but she could see all the way down both sides of the ridge. Virtually nowhere else on that hill would this be possible, and nothing about the place was evident from a distance. You had to get into her bed to observe the perfection. Only an animal who knows every inch of her environment could have found a place as fine as that.

The wildlife biologist Valerius Geist has said that on winter nights in the northern wilderness he has often taken advantage of the fact that deer know all the microclimates, some warmer than others. Where the deer kept warm, so could he. A swamp near our house creates a microclimate when its sun-warmed air drifts up a hillside. I learned about the microclimate from the tracks of the deer that use it in winter. Why so many tracks in this one place? Ah, yes, warm air is wafting by.

They have even found a sheltered microclimate under the branches of south-facing oaks and pines at the edge of a field. Most trees, being dark-colored, make little pockets of warmth around the base of their trunks, so that in winter they stand in their own little snow-free spaces, but in this particular microclimate, snow melts over a much larger area. There, the overarching branches protect the ground from snowfall but are high enough to admit the rays of the low, winter sun. Also the woods behind the trees are thick enough to keep the sun-warmed air from moving quickly through them. Thus the warm air stays more or less in place. Every year, even in the coldest weather, snow often melts under those trees, so the place is surely warmer than anywhere else, and deer lie down there to rest. Still another such place is at the top of the hill we live on, in the field north of our house. The deer who sleep there are those whose eyes shine in our headlights. After the snow melts and they leave their wintering areas, they sometimes go there just after dark. On warm nights, they sleep on top of the hill where the night wind cools them. On cold nights they sleep just over the crest of the hill on the side away from the wind. Always, they lie down at a certain distance from the woods so that any bad thing must cross the open field before it reaches them. This gives them time to get up and leave. In short, they have worked out every detail of their environment, and they use it to their advantage.

They know about us too. We are part of their environment, and just as they use the oak trees and the hemlock thickets, they sometimes use us as well. In spring a doe hides her fawns in our field. The grass is high by the middle of June, and for a number of reasons that include nesting bobolinks and monarch butterflies, we don't mow the field until the fall, so the grass is up to her belly by the time we notice her standing still, her ears a bit back and flicking, her tail also flicking, and one hind leg set wide. She is nursing her fawn. After a while she changes position and ducks her head. She is cleaning her fawn, removing excrement so the fawn won't have an odor. She doesn't like to keep her head down in the grass where she can't hear much or see anything, so she lifts it quickly every few seconds, has a look and a listen, decides she's safe, and lowers it again. When her fawn is settled, she walks back into the woods. The fawn lies down and waits. Or he doesn't lie down. He moves somewhere else and lies down. In a few hours, we see the doe again, looking here, looking there, until she finds the fawn. Again we see her standing, ears flicking, while he nurses.

All this takes place not far from the house. Whitetails are the world's most cautious animals. The woods are full of hiding places. What is there about our field that might induce a doe to hide her fawn there?

Oddly enough, our dogs might have something to do with it. We have always had dogs, and it's possible that the deer take advantage of their presence. Our dogs have always been shepherd types, not hunters, and although we have a cat who in his youth hunted deer, stalking them in the field and springing at them, always unsuccessfully, the dogs never try to catch them. Also, the deer can outrun our dogs. Thus, although the deer don't like these dogs, they feel relatively safe from them. If a deer in the field sees a dog by the house, the deer might snort and stamp a foot, but probably won't turn and run. One summer day I saw three female deer stare down a dog's relative, in this case a good-size coyote who had been resting in the shade of a boulder. He kept his head low, looking somewhere else while they were staring and stamping, and he got up and crept away when they advanced on him. One frosty morning I saw a large doe actually chasing a coyote, as if she meant to attack him. It would seem that adult deer are not panicked by small members of the dog family.

Very small fawns, on the other hand, would not be safe from any dog, but dogs seldom learn about them. This is because fawns are born without odor. Dogs have been known to pass within inches of hidden fawns without detecting them. Fawns are safe from our dogs because the dogs don't see them, don't hear them, and don't smell them.

However, if the dogs don't know about fawns, they do know about everything else, and consider it their duty to keep all comers at a distance. Our present dogs are little cattle dogs, both middle-aged females with Rubenesque figures. They are not formidable dogs, but when they see a stranger of any description, they rush at him, hair bristling, tails up, barking wildly. This keeps most people in their vehicles and most animals in the woods. The vigilance of our dogs has made a sanctuary of the field, and the deer take full advantage.

Once, a deer used us personally. My husband and I were standing outside when a doe came bounding out of the woods. Almost before we knew what was happening she changed course and ran straight for us, dashing by so close we felt the wind of her passing. Her mouth was wide open. We heard her gasp for breath. Then a wolf-size coyote ran out of the woods, right on her trail. He was less than a hundred feet behind her, also close to exhaustion but forcing himself onward. He stopped in his tracks when he saw us, though, gave us a deep, bitter look, then turned his head aside and grimaced, showing his teeth, as if saying to himself, *DAMN those people.* Then he slowly went back into the woods.

The doe had used us to scrape him off. We felt quite privileged. Not many people get to become the tool of a deer, or even to witness this useful strategy.

We are glad to be of use to any wild creature. Many years ago we planted an apple tree but have yet to eat a single apple. Always, others get there first. Wild grapevines grow over the bushes between our house and the road. A bear eats all the grapes, but we rarely know when he does this, as he comes very discretely. The difference between the natural and the unnatural seemed clearly understood by all concerned. Normally we wouldn't violate that understanding. But when we saw the miserable acorns in the fall of 2007, then saw how hard the winter promised to be, with an icy Halloween, a white Thanksgiving, and a Christmas so white we could barely see to get up our driveway, it seemed right to be offering a little corn.

Chapter Two

Cracking the Code

When I was a child and our farm was active, my friends were my younger brother and the kids on the next farm. All of us did farmwork, of course, but when our work was done we joined up to explore the woods and fields together, fascinated by the natural world, learning about it partly from direct observation, and partly from the adults of the local farm families. On our own, we acquired firsthand knowledge, but from the adults we acquired inaccurate lore. We countered some of the misinformation with a little observation—it wasn't true that porcupines throw their quills and snakes die only after sundown, for example—but because I was so hungry for knowledge and internalized every scrap that was offered to me, some of the lore stayed with me unnoticed unless I found a reason to doubt it. To my horror, I was in my twenties before I realized that there is no such bird as a chicken hawk. Eternally fascinated by the animal kingdom, I was ashamed I hadn't known that. Actually, I had known it. I knew that the birds we then called chicken hawks weren't always the same, but due to my early faith in grown-up information, it never crossed my mind that we were using one name for different species. I had swallowed this misinformation whole and unwittingly kept it.

That was why, when I began to feed the deer, I must have been blind to the relationship between them and other types of deer I had spent some time observing—red deer in Scotland and caribou on Baffin Island. We usually see the local deer one at a time—crossing the road, for example, or grazing in a field—so I assumed them to be essentially loners who got into your garden if they could, and lifted up their white tails when they ran. Farm lore held that they were venison waiting to be harvested. I was glad to feed them, but I didn't expect to learn much from them or about them because the farm lore that I unwittingly carried in my head told me that there wasn't much to know.

Even so, I thought if I was going to feed them I should make some effort to understand what I was doing, so I went to the literature. I found at least eighty books in print about whitetails, hundreds of other books that had gone out of print, thousands of articles in outdoor magazines, 1.4 million Web sites representing different levels of zoological expertise, and rafts of scientific papers. I discovered that whitetails are the most written-about mammals in the world and have been the most carefully studied. Certainly I didn't read all the books or visit all the Web sites but I did consult more than enough to safely say that a great many other people seem to agree with my early impression that whitetail deer are little more than potential venison. Much of what I read had to do with male whitetails and hunting—antler growth, fights, the rut or mating season—also population management, habitat, physiological facts such as weight loss in winter, and questions of appearance such as seasonal coat color and the like. Many of the Web sites are how-to postings about hunting, and even the field biologists seemed to be hunters at heart or hunters gone wrong, out there looking for data instead of meat.

I found exceptions, especially Dr. Leonard Lee Rue III, whose fascinating work dominates deer literature. He is a hunter, of course, surely a very good one, and he points out that his doctorate is honorary, that he was not formally trained as a biologist. Even so, he is entirely familiar with the abundant biological research. But the most important, most interesting information he presents to his readers comes from thousands of careful observations that he made in the woods, many of which are captured in photographs. There are things in the woods not found in books. His work is wildlife science at its best—a treasure trove of valuable, firsthand information. As the reader will note, I am so taken with his work that I cite him endlessly. The same cannot be said of many other writers who seem unaware that deer are motivated by anything more than instinct: "We know that a deer's brain is small and reasoning is impossible,"[1] writes one author, although the deer whom I encountered seemed to reason quite successfully, as do countless other animals. Many game managers don't want to know about that. To some of them, deer seem important only as potential victims. When searching the literature, I often felt I was eavesdropping on love letters between game managers and hunters.

So imagine my surprise, during the first few weeks of deer feeding, to learn not from the books but from the deer that they are social! Evidently the reading I had done was simply a grown-up version of farm lore. Well, that's not quite fair, because much of it was science, but the science of what? Not about what deer did with their time or how they related to one another, which, incidentally, was fairly easy to determine once I got started.

The feeding area was about a half an acre in all, bounded on the south and east by low stone walls with fields all around. It was also fairly near the house, as I didn't want to trudge long distances through deep snow carrying heavy pails of corn. Thus when the deer were eating they were close by, and I could watch them from a window.

The deer came every day, and at first, they all came at the same time. I'd notice them in the woods, sometimes in late morning, sometimes in mid-afternoon, coming along in single file with the biggest breaking trail through the deep snow. At the edge of the field the leading deer would stop for a look, then forge through the snow up the hill to the feeding area with the others close behind. Again the leading deer would stop and check for danger. At last it would enter the area. Meanwhile, other groups of deer would be approaching from all directions, stopping at the edge of the woods, thinking about coming nearer. The deer in the feeding area would throw up their heads and stand stiffly, necks long, heads high, ears forward, looking at the newcomers. The approaching deer would notice the stiff postures. They might wait uncomfortably, shifting around a little, until the watching deer relaxed. They then might cross the field quite quickly, sometimes on paths plowed by the earlier arrivals.

Sometimes the smaller deer ran ahead, as if excited by the prospect of corn. To me, this was exceptionally moving—hungry young deer who were born the previous spring and never before had experienced a winter, hurrying eagerly toward the corn. Soon, seventeen to twenty deer would be at the feeding area, some of them inside eating, others outside watching. Eventually, a few deer who had been watching might begin to ease themselves into the feeding area. A deer already inside might fix a hard stare upon a newcomer or might even rear up and strike at her with a forefoot. The newcomer might then take a few steps back but would not retaliate, and in time would try again, sometimes successfully. Taking the success as a signal, other waiting deer might then enter the feeding area and try to eat. If they were chased away from food, they might stay within the feeding area, waiting for a deer who was eating to move aside, but they might also retreat to the far side of the stone wall, whether or not they had eaten. There they would continue to watch.

At the edge of the woods, the flock of fifty-odd turkeys would take the presence of deer as a safety signal, and they too would start off for the feeding area, the leader plowing along in snow almost up to her neck, the others walking more easily behind her. I wondered why they didn't fly over the snow rather than plowing through it, then realized that as winter progresses and every calorie counts, turkeys normally fly *from* places, not *to* places, except at dusk when they fly to their roosts in the trees. In winter, they, like the deer, must not waste energy. Perhaps struggling through snow, however difficult, was nevertheless more conservative than flying.

When fifty or more turkeys were in the feeding area, they might begin to feel crowded, and like the deer, some of them might try to discourage others of low rank. This wrung my heart. The excluded turkeys would stand alone in the snow, looking at the fifty other turkeys who were eating and were welcome. So at the beginning of my feeding effort, I poured the corn in a very long line to try to accommodate every turkey. But soon I noticed that certain deer were taking possession of the entire supply, or trying to. After that, I put the corn in many separate piles placed far apart, so that everyone could have some, including the turkeys. This helped but didn't eliminate the problem, at least not for the deer. Some of them still managed to prevent others from eating, if not as often or as easily.

A deer would approach a pile of corn to nibble a little at the edges where the corn was spread thin, then move to another pile and nibble there—the way in which deer browse on plants. When certain plants realize that someone is eating them, they put out a bad-tasting toxin. Thus it's best to browse a little from one plant and then move on to another before the plant catches on. The corn, of course, wasn't exuding a toxin or changing flavor, but for all the deer knew, it might, so they kept to their standard practice. However, when a deer moved on to another pile, a deer who had been excluded might then nibble at the pile left behind. Since the piles were widely spaced, this gave most of the excluded deer a chance at some nourishment.

After about twenty minutes in the feeding area, one of the deer would flick its ears and whisk its tail a bit, as if thinking of moving on. Soon enough it would start slowly into the field. Others would follow. Surprisingly, even those who had not eaten would follow. Since the reason they hadn't eaten was because others had prevented them, I wondered why they didn't stay, to take a turn in the feeding area when they had it to themselves. There was plenty of corn.

But they didn't, and to me this seemed important. Very often, when an animal seems to be acting against its own best interests, it is because a more dominant animal wants it that way. Some animals will not eat in the presence of those who outrank them, as if to do so would be breaking the rules, asking for trouble in the form of a fight, or consigning themselves to ostracism. If deer were the loners I had thought them to be, they wouldn't do this. But there they were, doing it.

That they arrived together, interesting as this was, could have several explanations—safety in numbers, warmest time of day, former habits, and the like. But that they left together, even those who were hungry but hadn't eaten, has only one explanation, a psychological one. The turkeys did the same, and turkeys are nothing if not social. If ever there was evidence that deer are social, it was this.

Fascinating as that was, I knew I would understand nothing until I knew who these deer were. I had long since lost all hope of recognizing individual turkeys—there were just too many of them—but I did hope to be able to recognize some of the deer. So I got out my binoculars and began to stare at them, searching for features that I could recognize later. I wasn't completely new to the art, having been fortunate enough to have participated in field studies of animals in which recognizing individuals was essential. But I found this almost as difficult with deer as it was with turkeys. However carefully I examined a deer, he or she continued to look just like the others. Of course, they were different sizes, but I found it hard to tell just what size any deer was unless it was standing by another deer or by a tree on which I'd made a mark. And even then, what had I learned? That one deer could be bigger than another? I knew that already.

Color didn't help either, because as far as I could tell, the deer were pretty much the same color, dark in their winter hair. Even the fawns had lost their spots and were dark like the adults. Many deer have white rings around their eyes, some very pronounced, some with more white above the eye than below, some with rings so thin or faint as to be almost unnoticeable, and at first, I had high hopes for recognizing deer by their eye rings. But when six or eight deer arrived together, sorting them out by their eye rings was beyond my ability. Two of the deer had big white rings, and I could usually recognize these, but then, a few others had rings almost as big, and unless I saw these deer together to compare them, I got mixed up.

I could scarcely tell males from females. Antlers would have helped, but the males lose their antlers in winter. Still, some males had faint little whirls of hair where the antlers would grow in spring, and sometimes I thought I could see these with binoculars. Male sex organs weren't visible, or not in winter, as the testicles are inside the body and the penis is hidden in winter hair. Not even the manner of peeing was definitive. Females partially squat, as one might expect, but males also partially squat if they wish to pee on their feet. Why would they want to do this? Here, the literature was helpful. According to the experts, they want to soak the hair on their tarsal glands with urine. Deer are olfactory animals and the practice creates a meaningful scent.

I all but gave up trying to detect the males, but I did think I might be able to identify females if they were running away with their tails up. I used binoculars to search for a vulva, and sometimes thought I saw one. But unless the doe is in estrus (which they aren't in winter), her vulva is small and closed and is more or less hidden by the white hair of her rump patch. And by the time I'd seen the vulva (assuming I really did see one) the deer was too far away to transmit further information.

Determining gender by behavior was possible but prone to error. The males were more careless than the females. Whether this was because they were more confident, or because they were less concerned with the safety of others, I didn't know. If startled by something, they might simply run away with their tails down or lifted loosely, not raised up high and flagging, like the females. The females were more cautious, looked up more often, listened more often, and if alarmed, immediately raised their tails in a yard-long display of pure white hair—the white tail and white rump patch, one above the other—with the tail vigorously waving side to side. This must be the most conspicuous sight in all of nature. No deer can miss it, none can ignore it, and all who see it immediately bound away with their own tails up and waving.

Unusual markings were helpful, but I found very

few. Two young deer had ridges of hair on their necks, almost like little manes. Another had a tuft of hair on its back, as if from an old injury. Still another—a fawn—was exceptionally fuzzy like a little buffalo. But most deer didn't have unusual features. Sometimes they seemed to, and this was deceptive. For instance, all deer have small white spots inside and outside their hind legs and between their toes—the marks of different kinds of scent glands—and when I saw a large deer with enormous white spots on his legs, I thought I had him. But when I saw him again without large spots, I took him for a different deer. It *was* him, but when I first saw him he had spread open the patches of hair that covered his tarsal glands, probably because he was near a human habitation and was nervous, and was broadcasting his concern with his tarsal pheromone. The next time I saw him, he may have felt more secure, because he kept his tarsal patches shut like the others.

I seemed to be getting nowhere. Then one day, it came to me that my thinking was all wrong. I was viewing the deer as an aggregate of individuals. But deer are no more that than we are. I may be an individual, yes, but only in a way. Otherwise, I'm my husband's wife, my children's mother, and my grandchildren's grandmother, and thus am considerably more than just an out-of-context member of my species. As such, I'd be hard to locate in a crowd. The observer would need to have learned my various features, just as I was trying to learn those of the deer. If he then saw another woman about five feet two with short gray hair (how many of *those* could there possibly be?) he could get us mixed up.

But together with my family I'd be easy to spot. Reliably, our group would have the same number of big ones, middle-size ones, and small ones every time. We would come as a group and leave as a group, although we might mix with others when we got there. But once an observer had identified our group, he could then note a few special characteristics of some of us, maybe a mustache on the oldest male, two young females the same size, and a tall, dark-haired female with a little child, and he'd have us nailed. If it was me he was looking for, he could scan the group for the gray-haired female, and there I'd be—the only one. What's more, any such group would be different enough from any other group that if someone was missing or a visitor was added, the group would still be recognizable, and the change could be noted. None of this might be true a year later—some of us might be bigger, another infant might be present, and still others might be permanently missing, but we are not speaking of a year, we are speaking of a season. The same is true of deer.

This was the breakthrough. From then on, recognizing the deer was almost easy. Of course, since they were deer, not primates, their groups were not like those we might form but were appropriate to their own kind—the females close together with their young, and the males associating loosely—a modest version of deer such as red deer and caribou, who form large, highly organized groups of mothers, daughters, and children of both genders with mature males scattered nearby. As for the whitetails I was watching, a group of males might have four members today and six tomorrow, or disintegrate completely as the former members went different ways, or a yearling male could join his mother briefly, giving her group a surprising extra member. But the female groups are fairly constant, so recognizing them even with their male associates became possible. There they were, day after day, pretty much the same in the spring as they had been all winter. Deer grow very little if at all in winter, so even the fawns seemed unchanging. For me, this was a great experience. It was something like learning a language that seems impenetrable at first, just a mixture of noise. But if you try hard enough and keep at it long enough, sooner or later every little word has meaning.

Chapter Three
Deer Families

The first group that I came to know was led by a large, very beautiful doe who traveled with a second, slightly smaller doe. Female deer stay with their mothers for indefinite periods, perhaps until they themselves have fawns and thus are fawns no longer, so this young female was undoubtedly the beautiful doe's daughter, possibly born in 2006, possibly earlier. Also with the beautiful doe were twin fawns, obviously hers, born in 2007. To help secure this group in my mind, I began with the letter *A* and called them Group Alpha.

As it turned out, the name fit them perfectly. The beautiful doe had status, lots of it, and, interestingly enough, so did her children. Her twin fawns were too young to exert much authority, but if the feeding area became crowded, her highborn older daughter might stare at a newcomer, or raise a forefoot as a threat, or even strike at the newcomer. Just a stare was usually enough to cause the newcomer to lower her ears, turn her head slowly, and back away. When the newcomer was at a safe distance, her ears would come wistfully forward and she would watch the princess eat. This was true, even though she might be bigger and thus older than the princess. On one occasion, the princess dared to threaten a large male who was preparing to jump over the stone wall into the feeding area. He didn't like the way the princess was looking at him, had second thoughts, and didn't jump. Mostly, though, the deer whom this privileged group excluded were thin or young or both, and thus were surely of low status. Blessed are the poor, but not in deer cosmology.

In normal times, foraging opportunities are spread widely enough so that most deer can nourish themselves. This, of course, is why low-status deer are present at all—the forest is too big for the high-status deer to control everything, and also, since there's enough for everybody, they don't need to. However, when forage is limited, the high-status deer get most of it. When, for example, the acorn crop is scanty, yet acorns are needed for weight gain for winter, the high-status deer get most of them. It's easy enough to control the space under an oak tree. Thus low-status deer can be thin when winter comes. The pregnant females can have trouble nourishing their unborn fawns. In spring, such fawns can be very small at birth, and they may, if they live, always be smaller than the highborn fawns. In contrast, the high-ranking mothers control the best foods, hence their offspring are better nourished during pregnancy, and are led to the best foods thereafter. When food is limited, especially in an area that's easy to control, the high-ranking deer can own it.

Darwin's "survival of the fittest" has been used to justify all manner of human indecency, and some people may find high status unfair or even unpleasant in animals. It didn't make me happy to see the more vulnerable deer sometimes excluded, but the deer felt otherwise, and took whatever measures they thought were appropriate to promote their own well-being and that of their children. Dr. Rue has this to say about such exclusion: "In the world of nature, there is no such thing as fair play; everything is done for keeps, everything possible is done just to survive."[1] All I could add to that is that deer have been around for five million years and must know what they're doing. As far as their population is concerned, it's good that some of them are in top shape, even if all of them cannot be. A winter such as that of 2007–2008 can bring a population to its knees. But the healthiest deer may survive. If so, a few years later, deer will be plentiful again, in all their social complexity. I suppose I could have tried to scare off the highborn deer when I saw the humble deer coming, but I didn't, not only because the humble deer would run away with the others, but also because the exclusions were due to deer politics, and were thus immune to human influence, just as our politics would be immune to the influence of deer.

Often while watching the highborn deer eat corn, the humble deer would eat twigs from the surrounding bushes. Sometimes a dry hydrangea blossom would blow off its bush and roll toward them like a little tumbleweed. They ate those too. The highborn deer didn't seem to care, so probably all concerned saw an important difference between the corn and the other, natural foods that were present. The corn had lots of calories, and the twigs and dry blossoms had very few, hence it was food value, not the act of eating, that was causing the distinction. It was as if the humble deer were saying to the highborn deer, "The better things are yours."

What gave Group Alpha their status? My best guess is that the beautiful doe was in the prime of her life, therefore deserving of status, and that many of the other deer in the area were her children and her children's children, and as such would acknowledge her importance. After I learned the identities of these groups, I realized that this beautiful doe was usually the first to come to the feeding area. She would lead her group down a hillside to the south, across the ice on a pond, and through our field toward us. Her approach was the signal for other deer in far-flung places to leave the woods. How did they know that Group Alpha was coming? It's hard to say, but it didn't seem to be because they saw them or heard them. As far as we know, deer do not make ultrasound or infrasound, hence if Group Alpha had made a noise that could be heard by deer half a mile to the north, I would have heard it too, and I heard nothing. Nor did the time of day influence their joint arrivals—all they seemed to need was strong daylight. My best guess is that the answer lies in smell. Deer are olfactory animals, and thus produce informational odors and pheromones that speak to others. Most deer came from the north and east, but the beautiful doe came from the south with the prevailing wind behind her. It could be that her signal was carried on the air.

She seemed to be much on the minds of the others, even when she wasn't present. On the rare occasions that others came toward the feeding area before she did, they would stand in the field and look around, perhaps to learn where she was. Or they would fix their eyes on the woods to the south, as if they were expecting her to emerge. They held their tails loose, twitching slightly, but they didn't stamp their feet or seem agitated as they would if they suspected predators. Perhaps they felt that they were trespassing and were nervous to be eating without permission.

This was surprising, considering that all were hungry. This was what told me that there was more to deer society than I had at first supposed. Similarly, the departure of the Alphas was also a sign to the others. They followed whether they had eaten or not, even though there was still plenty of food, even though the departing deer had no real way of knowing what would happen in the feeding area after they left. Day after day, the Alphas would leave the feeding area when they felt like it, cross the ice on the pond, and climb the next hill toward their shelter area, which I believed was in a massive growth of evergreens on the north side, the lee side, of the hill. The four Alphas would walk slowly, always in single file. Four or five others might trail directly behind. The rest of them would head off for their own shelter areas, and the feeding area would empty, like a theater after a performance.

Why did the other deer insist on coming when she did? Usually, the deer spent no more than half an hour in the feeding area. That left about eight hours of daylight during which anyone could have been there. Why didn't some of them take advantage? One can only guess, but since this happened in the early months of winter, when all the deer must have been uncertain about approaching a house, it might be that if a doe of her stature found the area acceptable, the others did too, and came because they respected her wisdom and because there was safety in numbers. That explanation seemed simple and obvious, but many things that animals do are neither simple nor obvious, and while the explanation may be somewhat valid, I didn't think it was enough.

I haven't always lived in New Hampshire. When I was in my late teens and early twenties I lived in the unexplored interior of the Kalahari Desert in Africa with people who were hunter-gatherers. My father was more than a farmer. He was also a civil engineer and a businessman, and later in life, an explorer. My mother was an anthropologist. That's why we went. The experience stayed with me, and ever since then, I've seen the world through the lens of the Kalahari. The people among whom we lived had been there for 35,000 years at least and as such, were living the life we all once lived when we were part of the African savannah fauna. They lived in the Old Way, and their culture had enormous stability, meaning that they kept the Old Rules that were laid down by necessity—the kind of rules that all who live in the natural world must keep. But unlike other species, ours can be interviewed, and today, when I find a mystery of nature, I turn to the discussions we had with the hunter-gatherers of the Kalahari for a possible answer.

So when I saw all deer depart my feeding area just because the Alpha group was leaving, I thought of the Kalahari, and remembered the respect its people had for other people's things. These were tied so closely to the owner as to be almost part of the owner. If, for instance, we Americans find a gold ring on the street, we pick it up. We might then keep it or might try to find whoever lost it. Either alternative is more or less acceptable to us. But if the Kalahari people found a similar object they'd leave it where it was. Why? Because it belonged to someone. That the owner wasn't with it didn't matter. He or she would probably come back for it someday. To violate the concept would have been to violate the all-important intergroup relations, and never happened. Our relationships and our social systems were our lifelines during the 150,000 years that our species lived on the savannah, just as they are for any social species. The Old Rules must be kept. If the deer at my feeding area departed when the Alpha doe departed, then perhaps it was an Old Rule they were keeping—the kind of rule that governs what you do, whether or not others are watching.

The theory of safety in numbers has merit too. I found some support for it in that when the deer came, the turkeys would take their presence as a safety signal and come too, and would mix right in with them. In order to eat, the deer would have to put their faces down among the turkeys, even touching them. Interestingly, some of the deer who chased other deer seemed flummoxed by the turkeys. Only rarely would a deer try to chase a turkey. When the turkeys finished eating, they would depart in single file without a backward glance, as if the deer no longer mattered. My guess that deer came together for their mutual safety was somewhat confirmed toward the end of winter. Then, deer seemed to come of their own initiative, as if they no longer needed proof that the area was safe. By the end of March, they were not coordinating but seemed to come and go in their small groups pretty much as they pleased, and even the humblest deer were eating.

It was hard to be sure, because in March, the beautiful doe stopped coming. I think this was because the ice was softening on the pond. As far as I know, the last deer to walk across the ice did so on March 16. I held my breath as I watched her, wondering what to do if she fell in. But the ice held, and she made it. A few days later the ice was almost gone and the brook was over its banks. The beautiful doe and her family would have had to add almost a mile to their journey, going upstream to a crossing place, then coming back downstream to reach my feeding area if they didn't want to swim or wade through deep, icy water. Perhaps they didn't want to spend energy by traveling so far. In theory, they could have come by road, but the deer of these parts don't travel along roads, although they cross them. However, the Alphas seemed in good condition when I last saw them, and snow on the fields was melting, so grass was already exposed. If anyone could make it through to summer, the Alphas could.

A deer with high status could be a detriment to some, but she also offered advantages, not only to her own group, but also to all others. The beautiful doe of Group Alpha was watchful and cautious, which, I think, was because of her responsibility. She kept watch more carefully than any other deer, and others trusted her messages. Not that they ignored warnings by others, but many of the younger deer could fling up their heads or flag their tails without causing general consternation. Sometimes a young deer in the feeding area would flag and dash off as a trick. A few others might take alarm and do the same, at which point the young deer would turn back and eat the food the others had abandoned. But the beautiful doe was far above that sort of trickery, nor did she need to resort to it—if she wanted someone else's food she had only to take it. She was in charge, and everyone knew it.

If she looked at something in the distance, others would move near her to try to see what it was. Sometimes she was watching the turkeys, gathering into their flock of fifty, preparing to march in single file across the field to visit the feeding area. More often she was watching other deer, perhaps four, perhaps six, deep in the woods but on their way to the field. If the beautiful doe moved out of the feeding area for a better look, those near her went with her, also for a better look. The approaching deer would stay in the woods until the beautiful doe stopped staring.

Like many other deer, she found the sight of midsize animals unpleasant, and if she saw one, perhaps our cat who had escaped from the house or a fox at the edge of the woods, she would snort and stamp one foot. Then she'd stamp the other foot. Once she stamped at me when she noticed me looking out a window. By deer standards, I would not be a midsize animal, but on that occasion only my head and shoulders were showing. Her object was to make sure that no danger would come to her followers. She would slowly approach the suspect, head high, eyes threatening. The deer around her might do the same, but they would make sure to keep the beautiful doe in their peripheral vision so they could see what she was doing. If the suspect was a distant fox, he might just keep going. The deer would watch him out of sight. If the suspect was our cat, she would not be pleased by the staring, and would go back into the house through the dog door. The deer would then continue eating.

When the beautiful doe was preparing to leave, she would move around a bit at first, looking here and there, flicking her tail as if she felt unsettled, perhaps because she was picturing the path she was about to take. Few things are predictable in the natural world. This is an Old Rule, and the deer know it. The thicket where you browsed while on your outward journey could hide a predator on your return. Everything can change in a split second, and every sound or scent, every flicker of movement that you catch in the corner of your eye, can tell you something. The Old Rule demands that you stay in the moment—always, always in the moment. So if the beautiful doe was nervous, she might have been preparing herself to guard her followers—her grown daughter, her little fawns, and perhaps six or seven others who might trail behind her rather than taking their own directions, in part because of her status, in part because of her ability, and in part because she surely owned the region's best shelter area and might let them take shelter with her. I'm not sure that she always did. No matter how many others followed her away, when she came back the next day, often just her daughter and fawns were with her. But not always. Sometimes an entourage followed her. No doubt they were her older children and her grandchildren.

One day while eating corn, the beautiful doe flung up her head and kept it up. All others noticed, and also looked up. She then pressed her lips tight together and drew an extremely long, deep breath, as if pulling in scent from far away. For a fraction of a second, she held her breath, as if considering the scent. Then she stamped her foot. Whatever she found in the air, she didn't like it. But she needed more information. So she walked toward the scent with great determination, then stopped and stared at the woods with her nose up, her neck stretched, her ears far forward, and one hind leg set outward, a posture deer assume when experiencing real concern. The others joined her until all were looking. She stamped again, and again, then suddenly she wheeled and bounded away in the opposite direction, her tail held high. The group around her exploded, and all went bounding after her, not in single file as deer often travel, but in a big, tail-waving mass. In seconds, all were gone. I grabbed the binoculars to try to see what had frightened her, but saw nothing. You didn't have to be a deer to know that something bad was there, though. The beautiful doe could see right through the woods, and smell what she was looking at. And whatever it was, it was she who had noticed. No wonder others relied on her.

Another group of five deer—two large and three small—was also easy to identify, or they were after I managed to distinguish them as a unit. Sometimes they were trailed by three others, making a group of eight, but the five were separate often enough to be a group of their own. These deer became Group Beta, and here again, the biggest doe was the mother. The second doe would have been her grown daughter. The fourth and fifth deer, the small ones, were her twin fawns born in 2007. But the third deer was a mystery. This one was bigger than the twins, and smaller than the grown daughter. To me, this said that the third deer wasn't anyone's littermate. Also, the third deer wasn't the same color as the others. They were dark in their winter hair. The third deer was conspicuously red, almost in summer color. Yet this mysterious youngster was strongly a member of the mother's group, right in tight with the others.

There seemed to be two possibilities. Maybe Deer

Three was male, the triplet of the two small fawns. Male fawns are substantially bigger than female fawns at birth, and stay bigger throughout their lives. Sometimes Deer Three acted like a male, chasing other deer, for instance, sometimes striking at them with a front foot, which might have been a gesture, except that Deer Three sometimes hit the victim on the back, which hurt. Sometimes Deer Three reared up and struck at another deer with both front feet—right-left, right-left, right-left—in rapid hammer blows like a boxer. The blows didn't connect but were threatening. Both males and females do these things of course, but all in all, for one so young, Deer Three of the Betas was the most aggressive deer to visit the feeding area, and others often moved away when they saw this youngster coming. Deer Three was also bold, and might run ahead of the group as they walked around in the field, often far ahead of the others. Deer Three would even lead the group as they went to investigate something worrisome, such as the turkeys as they emerged from the woods or our cat, who once again had escaped the house and was sharpening her claws on a tree.

The mother of the Betas didn't seem to have quite as much status as the beautiful doe of the Alphas, but she was highborn nevertheless and seemed serene about it. Perhaps her lawless little fawn had taken this as encouragement. Yet I couldn't help but feel that something was a bit off. In keeping with the customs of many other ungulates, most young males the size of Deer Three form loose associations, at least for short periods. An aggregate of six young males—the Rho Group, four of them the same size as Deer Three—would have been perfect company. But no. Deer Three stayed with the mother.

Yet this was not conclusive proof of gender, so I searched carefully with binoculars every time I saw this third deer, looking for the whirls of hair that predict antlers. I didn't find them. True, these whirls are very faint and often don't develop until later in life, and I am sure I have missed seeing them on many a forehead. Still, the forehead of the third young deer was as smooth as silk, a doe's forehead. That didn't mean that Deer Three was a doe, but it made me wonder. Might Deer Three be unrelated to the others? Might that explain the red summerlike coat so different from the others? If so, could Deer Three have been adopted?

Very occasionally, a doe will adopt an orphaned fawn even if she has fawns of her own. So it is possible, however unlikely, that the leader of the Beta Group adopted the third deer. In the summer of 2007, a doe was killed by a car not far from our house. Possibly she was the birth mother. If this were the case, an orphaned youngster, probably in need and unattached, had been allowed to join the Betas.

The leader of the Betas was a very nice deer. She seemed kind, unlike some others. When her group came into the field, they walked in an easy, almost casual manner, showing few signs of stress or strain, although taking all the usual precautions. Rather than follow in single file, all the young deer might cluster around the mother, as if that was where they want to be. Now and then, the twins would frolic. Now and then, they would touch noses with their mother. She seemed to be very present to all three of the young ones, gently aware of them, as if she welcomed them beside her.

One day as her group neared the feeding area, Deer Three chased one of the twins, who ran behind the mother. The mother quietly raised a forefoot and gently touched the aggressive youngster under the chin—a strike gesture, but greatly modified. The youngster wavered slightly and seemed a bit humbled, and then left the twin alone for a while, as if in respect for the mother.

I saw the Beta group often. They came and went from woods to the east of the field, and I think they sheltered on the side of the ridge on top of which, a few years earlier, I had surprised a doe in her bed under a hemlock. Perhaps the doe I surprised was the mother of the Betas, when she herself was young. That slope is perfect for deer. It's out of the wind with thick evergreens for shelter, and is right above a little swamp from which faintly warm air sometimes rises and cold air drains down and away. It's like the hill I live on, which was a farm long before my father revived it. On my hill, too, cold air drains away from us, and windborne frosts hit above us on the mountain. We are the last to lose our garden due to frost. This was the wisdom of the old farmers—before they chose the sites for their farms, they found out where the deer stayed. My hill is so perfect for deer that they still use it at night when we're asleep. The next ridge to the east is no different, except that its slopes are forested. It made me happy to think that this interesting family of Group Beta might stay there.

Mostly the Betas were a group of five, but often enough they seemed to be a group of eight. This was because of the Deltas—a doe and her two daughters who sometimes trailed the Betas and sometimes may have shared their shelter. However, their more permanent place seemed to be to the southeast near the large swamp. Possibly they sheltered in the warmer microclimate above it. The only disadvantage of that microclimate was that most of the trees were hardwood and did not offer the protection of an evergreen grove. There were some evergreens, but not many. If the place were better, Group Alpha would have owned it. The Deltas might also have sheltered in evergreen brush at the edge of the swamp.

The two Delta daughters were different sizes, which meant they were not twins. Yet they were almost the same size, which was confusing. They were always with their mother, and always ate what and where she ate, so a difference in nourishment didn't seem to explain this. They were also quite different in appearance. The smaller one had a long, pointed face, and was very graceful, even for a deer. The larger one was just like the mother, if not quite as big. One day it came to me that the birth dates of the daughters might be 2005 and 2006, as female deer stay with their mothers for undetermined amounts of time. It also meant that none of them, including the mother, had a surviving female fawn from 2007. All three were unassuming deer, coming and going without much fuss, minding their own business, often joining the Betas at the edge of the field to approach the feeding area, often waiting for the Beta doe to enter first, then following close behind her. I then began to wonder if the Beta doe might be the Delta doe's mother. That would be why the two groups merged so often. Since most groups of deer, although almost certainly related, seemed nevertheless to be distinct and separate, I attributed their closeness to the kindness of the Beta mother.

If during that winter I was asked which deer seemed most significant, I would never have mentioned the Deltas. Yet it was these deer whom I was able to watch the longest. When spring came, they were still with the Betas, or more or less with the Betas. But by the middle of June, the Betas had vanished, all five of them. Deer have winter ranges and summer ranges, and since it didn't seem possible that something bad had happened to all five of the Betas all at once, I assumed that they either went higher on the mountain or else went to the woods at the west of the road, a popular deer habitat.

But the Deltas stayed put. Right through the month of June I saw them every evening just at sunset, calmly grazing in the field, their dark winter fur long gone, their red summer fur shining in the grass. I hoped to learn if any of them bore a fawn. All of them seemed in good condition and likely to be able to carry a fawn, especially the mother. If I had helped to enable this by feeding them, I was happy.

One afternoon in January after many heavy storms, a new group of five deer appeared at the feeding area. Every deer was looking thin by January of that winter, but the biggest of these four—a tall, dark-colored doe—seemed thinner than most, her neck shrunken, her hip bones showing under her skin. She stayed the longest, eating, eating, even after the others heard an approaching motor and went into the field. Thus the tall doe was last to leave when the snowplow came up the driveway. But, of the five, she ran the fastest, probably because she was vulnerable. The local predators take the slowest, weakest deer. Perhaps she pictured a predator in the woods, watching her departure. *No use chasing me,* her manner said. *You won't catch me.*

These deer became the Tau Group. I was very moved by them. With the tall doe traveled two grown daughters, evidently twins, and two very small fawns of the spring of 2007. These fawns were by far the smallest deer who came to the feeding area, and were dangerously small to survive a New England winter. But they didn't seem to know how fragile they were, or what hunger and the winter storms would do to them. On their first visit, they played together, frolicking around behind their elders as if all deer lived forever. Later in winter, hunger and the cold were telling on them. They seemed tired, moved more slowly, and never played or frolicked.

Why were they so small? Perhaps they had been born late in the year, suggesting possible problems with their thin, dark mother, such as delayed estrus. Or perhaps their mother had not eaten much during the past winter when she was carrying them, so they might have been undernourished in her womb and then were born tiny. All this suggests poverty. The Tau Group was humble. They knew they weren't important. Sometimes they would come right up to the edge of the feeding area, but then stand there together and watch the others eat. Sometimes the fawns would try to make their way to some of the corn, but other deer would chase them. After that they would stand behind their elders and watch. Even then, they looked at the food with ears up and eyes wide, as if they still were hopeful.

As far as I know, I saw the Tau Group about eight times. On their first visit, only the hungry mother ate. On the second visit, they all had an opportunity to eat. On the third visit, the grown daughters ate while the mother and the fawns looked longingly from the sidelines. Around this time I began to wonder why, if the other deer worried them, they didn't come at night, when the others weren't looking. Food was available, I saw to that. But then I realized how bitterly cold it was at night, below zero, and colder still with a strong wind blowing. Unless these deer sheltered very nearby, which they didn't, the journey would have cost them far too much. The twin fawns in their tiny bodies could have lost what little warmth they had, and died from exposure. However, this also raised another important question, which has no easy answer. Since they lived fairly far away—half a mile to a mile, as I believe I discovered—how did they know about the corn? I'm not sure there's an answer for this, at least not one that a person could fathom.

I know the Tau deer lived through most of the winter because I saw them early in March, one afternoon as I was driving to town. I was half a mile north of my house when the leader of the Tau Group crossed the road in front of me, going south. Back in the woods was a low, forested area much favored by deer, and I wondered if she sheltered in there. That could explain why we saw her so seldom and also why others prevented her from eating. Perhaps she lived too far away to belong among the others. This, of course, is why the experts tell you not to feed deer. My corn was luring this vulnerable family away from their winter shelter. But the Tau mother was so thin and her fawns were so small that she perhaps did what I would have done with my children under similar desperate circumstances—risked everything on the chance that some of us would live.

It was a sunny afternoon, about 20°F after a very cold period. Perhaps she was taking advantage of better weather for a foraging journey and might have been heading for my house. I looked around for the rest of her group. They were in the woods at the roadside, about to follow her. Then a strange thing happened. As any deer would do, the two older daughters began to turn aside, as if to run away, but then they didn't, and instead they looked right at me. So did the fawns. They looked and looked, right into my eyes. I thought they recognized me. They might easily have recognized my scent, which would be wherever the corn was, and they could have known the scent and appearance of my car, which was usually nearby when they came. Perhaps one of these features told them something. Perhaps the feature spoke of food, and of living through the hardest time of winter. At any rate, they looked at me until I worried that I was keeping them from their leader, separating their group, and I didn't want to do that, so I drove away. I don't know if they came to the feeding area that day. I was on my way elsewhere and had to keep going.

I also don't know what became of the Tau group. After that, I saw only the mother, one of the grown daughters, and one of the fawns. It seemed only too likely that the others didn't make it, surviving the worst part of winter only to die in early spring. The survivors stopped visiting my field and feeding area before I stopped putting out corn, but by then a certain amount of browse was available, so I told myself that these three, at least, were still alive, if elsewhere.

Why did I feed these animals against all advice? Because we live in the same place, because they were individuals, because they had relatives, experience, a past, and desires, because they were cold and hungry, because they hadn't found enough to eat in the fall, because each had just one life. One spring before I began feeding the deer, after another terrible winter when our house was dark in the daytime because snow had drifted over our windows, I found three deer carcasses near the house—their long winter hair spread in characteristic circles about four feet wide with a few bones in the middle. One was very near the house—not thirty feet away on the south side of a stone wall. Perhaps coyotes came upon her. Or perhaps she simply lay down in the shelter of the wall and died there from starvation and exposure. After that, perhaps coyotes found her. By morning, snow had covered all traces of this very sad event, or I would have seen something. I did not report these carcasses to the game warden (only because I didn't think to do so), but every year such carcasses are reported, and when the bones are examined, the marrow is red instead of white. Healthy marrow looks like lard or suet. Red marrow means that every last gram of fat has been used up. Deer with red marrow are too weak to keep going. They lie down and die of hunger and cold.

Once in 2007 a blizzard began early in the morning and lasted all day, bringing over a foot of snow. The turkeys were miserable. When they tried to walk, they would sink in, then would try to climb up on top of the snow, only to take a few more steps and sink in again, a constant, exhausting, calorie-expensive struggle. They seemed to have little hope of getting to the feeding area, so they gathered loosely in the middle of the field where snow began to cover them. In the early afternoon they decided to roost, hours before they normally would do so.

In less severe weather, they would gather in the field at dusk, a few hundred feet from the edge of the woods, forming a pear-shaped flock with leaders at the front—the stem end of the pear—where they would look at the trees for a long time, as if picking out roosting sites. Perhaps they didn't want to fly among close branches with no plan in mind. At last one of the leaders would start to run, then launch into the air, then fly in a long, low trajectory to the branch of her choice. After a time, another would fly, then another and another until all were high in the trees. Usually the process took about ten minutes, from forming the pear until all were roosting, but on the day of the terrible blizzard they stood facing the woods for a very long time. The air was thick with whirling snow. Perhaps it was hard to pick out a branch in such low visibility.

At last, one of them began the roosting process, lumbering forward a few steps through the deep snow, then launching into the air, flapping hard but flying heavily, with dozens of wing beats per yard of altitude. When that first turkey was in a tree, another turkey launched, and then another, struggling for altitude as the snow on their wings weighed them down. Why they roosted one at a time, I didn't know. They always did, though, but on the day of the terrible blizzard the process took them almost half an hour and not all of them made it. Those who did tried to roost near the trunk of a tree in hope of a little shelter, and for the rest of the day and all that night they clung to their branches with their heads under their wings while the blowing snow covered them. Those who could not fly crept in among bushes.

A number of deer came several times to dig for the corn, despite the whirling snow. If possible, it is always best to eat before the snow gets too deep—you never know how long the storm will last or if you'll be able to move around later. Nine deer were there, and just before dark they were approached by three others. I didn't know them. I could barely see any of the deer through the whiteout, but I noticed that the newcomers were wary of those who were eating and stood at a distance, watching them. When the other deer left for the woods, these three left too, but went to a little island of oak trees in the middle of the field. Instead of moving into the woods to curl up in the shelter of low-growing evergreens as the other deer had surely done, they stood exposed among the tall trees, getting colder and colder as the wind blew snow around them. They were standing there when it got dark.

That night, because of the blizzard, I didn't put out corn as usual. I thought the wind would blow it away, and that I should wait until morning. Late that night the storm seemed so violent, with wind plastering snow against the windows and branches falling, that I stepped outside to try to judge the damage. The snow was deep all around the house except where the deer had been eating. There, despite the whiteout blizzard, I saw three enormous circles where much of the snow had been cleared. Then the wind sucked backward for a moment, revealing the three deer who had waited on the island of trees. They didn't see me. They seemed desperate. They had come back for what little was left of the corn, when other deer were not watching. There they were, white with snow—even their eyelashes heavy with snow—digging, digging.

They wanted to live. That's why I fed them.

Chapter Four

The Hazards of Feeding

We are told that deer-feeding is unnatural, but is this true? I was heartened to learn of another deer/primate alliance, one that is out of human control and beyond human comment. In New Hampshire, of course, the deer are whitetails and the primates are people. In Central India, however, the deer are chitals and the primates are langur monkeys. Troops of langurs visit certain trees to pick the leaves. The langurs eat only the stalk of the leaf and drop the rest on the ground. Chital deer gather in large numbers under the trees to eat what the langurs drop, just as deer gather in large numbers in my feeding area. The chitals jostle or kick at one another, and the high-ranking chitals try to exclude the low-ranking chitals, not unlike the whitetails who visit me.

The leaves provided by the langurs are important to chitals in the dry season, which in a way is like winter in New Hampshire. In both cases, food is scarce, consisting mostly of dry vegetation that lacks nourishment. But the leaves remain succulent, and a troupe of langurs might drop about one and a half tons of them per season, which is probably more than the amount of corn that any deer-feeder might distribute during a New Hampshire winter.

The most famous alliance between primates and deer is that of the Lapp or Sami people and their reindeer. However, in this alliance the reindeer derive no benefit. They find their own food in the woods until the herders round them up to exploit them, using them for meat, clothing, and labor. The reindeer have enormous spiritual significance for the herders, but in biological terms—and there's no polite way to say this—the herders are parasites of the reindeer. The herders derive enormous benefit from the relationship, and the reindeer derive none.

In contrast, in the alliances formed by chitals and whitetails, the deer derive significant benefit. They get life-giving food that apparently is thrown away by the primates. One might say that they alone derive a benefit. But that is not quite true. Their relationship with langurs, in biological terms, would be described as "asymmetrically symbiotic," meaning that both derive benefit, but not equally. The deer get food, and thus benefit the most, but the primates get important information. The langurs learn of nearby tigers or leopards from the chitals, who, like whitetails, have marvelous olfactory abilities. In my own case, I learned enough to write this book.

Thus deer-feeding has precedence in nature. Even so, the New Hampshire Fish and Game Department in cooperation with the University of New Hampshire Forestry and Wildlife Program puts out a flyer entitled "More Harm Than Good" to discourage people from feeding deer, claiming it may kill them and listing the damage it inflicts, even if it doesn't kill them. This publication must be taken seriously. I can easily understand the need for the flyer, even though Fish and Game contradicts itself in another publication, the annual *New Hampshire Hunting Digest*. There, they say that feeding deer allows too many to survive, which causes overpopulation. Either argument could be valid—the unnatural is never good—and I would not have fed deer if, in the fall of 2007, the oak trees had done their part, which they didn't.

The woman from whom I buy corn told me she had about one hundred other customers who also bought corn. Surely some of us were feeding geese or turkeys, but most of us, it seems, were feeding deer, and anyway, no matter what animals we intended to feed with our corn, deer were eating it. Each winter month, from November through March, we deer-feeders bought almost eight tons of corn and other grains, she told me. That's forty tons for the season. And we weren't coming from distant communities—we were mostly locals. This meant we were probably distributing our bounty within twenty or thirty square miles. Who knew what *that* was doing to the ecosystem? And the problem was not just ours—it was pervasive. Throughout the state in winter, game wardens try to monitor the condition of the deer by examining their droppings, and I've been told on good authority that almost all droppings have corn in them, no matter where they are found. This information, which I didn't learn until that summer, changed the nature of my guilt about wrongly interfering with nature. No longer did I see myself as a lonely giant of malfeasance, just as a tiny part of a general malfeasance. My crime seemed less like grand larceny, more like parking at an expired meter.

I tried my best to address the concerns expressed by Fish and Game and did all I could to mitigate the problems relevant to my situation. To start with, Fish and Game points out that a supply of artificial food can cause too many deer to gather in one place. This is perfectly true, just as leaves dropped by langurs attract too many chitals to one place, but I'm not sure this happened in my area, perhaps because, each winter month, eight tons of food were being distributed elsewhere. Also, thanks to earlier hard winters and one disastrous hunt in 2006 by a large group of trespassing men from a nearby town who managed to drive, encircle, and slaughter many deer—allegedly almost forty of them—the deer population of our area was not at its optimal level.

An added concern about deer in high concentrations is that they become aggressive, as the chitals in Central India unfailingly demonstrated. And yes, the deer in my feeding area were sometimes aggressive. Status was surely the main reason, but another is that with the exception of mothers and fawns, deer don't like to be too close together. When adult deer are eating their normal, wild foods they like to keep a certain distance. If they're grazing in a field, for instance, they often like to be in a line, with everyone facing forward and spaced three or four deer-lengths apart. For them, this is comfort, and they then eat peacefully, their tails occasionally flicking. Big as my feeding area was, it wasn't so big it could contain seventeen to twenty deer spaced that widely. This alone could have made them uncomfortable.

However, deer don't need artificial situations to make them aggressive. If two groups of deer are near each other in a field, even if miles of forest are stretching out on all sides, they still may chase or strike at each other. To them, status is territory, and territory is all-important. It is food and life. It is fawns that survive the winter. Even so, because injury is detrimental to any wild animal, most of them modify their aggression for this very good reason, and the deer in my feeding area were no exception. I watched all aggressive incidents quite closely, but never saw anyone get hurt—certainly not, as Fish and Game suggests, in fights that led to death or serious injury, although I'm sure this could happen, especially if two equally important groups from different areas came face-to-face with each other. In my feeding area, most of the deer seemed to belong to the same social order, and knew their places in it. Not one deer was injured by anything, let alone by the aggression of others, especially since most aggression was limited to staring, or to walking stiffly forward with flattened ears. The feeding area was large enough and the piles of corn were plentiful enough so that serious fights were unnecessary. The low-ranking deer didn't seem to want the risk, so they gave way when challenged. But I think that depended on their condition. Late in the winter, when all were stressed by hunger, some of them, such as the three deer in the snowstorm, came back to eat after their challengers were gone.

As for too many deer, once in December, in a very hard part of winter, twenty deer came, but only seventeen or fewer came each day thereafter, until late March, when the rules changed, and more deer appeared. As for the twenty dwindling to seventeen, what happened to three of them? I didn't know and couldn't even be sure that there were only seventeen, because in December I wasn't very competent at recognizing deer individually. If three had died, one would expect that the smallest, most vulnerable fawns would be the victims, but as far as I could tell, when the various groups of females came, they had the right number of fawns with them. Not until March 21, the so-called first day of spring, was I afraid that certain deer were truly missing when I thought I saw the mother of the Tau Group alone in our field. I feared for the others. Meanwhile, it became my impression that the three absent deer were those who had waited on the island of trees in the field during one of the blizzards. I wasn't sure just who they were, but they hadn't seemed to belong among the others. They reappeared in March, though, and after that were frequent visitors until the first week of April. By then, their antlers were budding. They were young males.

This brought the number of regular visitors back to twenty, or six less than are shown on the cover of the Fish and Game brochure, which shows twenty-six, with perhaps more in the wings. Are twenty-six deer together too many? Possibly, but on March 22, 2008, in an enormous field about five miles away, some people saw seventy or more deer together, an astonishing spectacle that had nothing to do with corn. Because much of the snow had happened to melt that morning, these deer may have been after the newly exposed grass. Many deer came to our fields on that same day, also to eat newly exposed grass, preferring it to grass that had been visible earlier. Who can say what is the right number, or the wrong number, of deer? As for our area, in the year we planted oats, forty-five deer showed up to eat them as they sprouted, coming back every day for three days until every last sprout had been pulled up and devoured. But during the winter of 2007–2008 I never once saw anything resembling forty-five deer together, let alone seventy. It would seem that I was feeding a fairly local population.

Fish and Game points out that a supply of corn can reduce the fat reserves of deer because it causes them to travel away from their shelters, and in winter, shelter is more important than food for survival. This is a very significant concern, especially for the fawns. With much smaller bodies than the adults, the fawns of the past spring are the most vulnerable to cold and exposure, but they are no longer being treated as infants. A doe who would hide and protect her fawn in spring would expect him to follow her in the fall, and probably would not stay in a sheltered place just because he was young, tiny, and cold. After all, it is something of a law of Nature that those of us who are old enough and strong enough to breed should stay alive to do so, even at the cost of our current offspring. They can't live without us, but we *can* live without them, so our responsibility is to save ourselves and try for more offspring later. Thus if an adult doe thought she needed to travel for food to save her own life, she'd do so, and her fawn would follow.

However, with the exception of the Tau Group, I do not believe that fawns were seriously endangered by my deer-feeding. I base my belief on the fact that several excellent wintering areas for deer would be, if not for the trees, almost within sight of my house. Several more were farther off in the woods. Two of these were deer yards where deer gather when the snow is deep, making a network of trails to trees where they browse. Other winter areas were simply thick patches of evergreens beneath which a few deer scrape away the snow and curl up where they are somewhat sheltered.

I learned of such places from my father, who took me to see deer yards when I was a child. For anyone who walks in deer-inhabited woods, deer yards may be a common winter sight—if you find a sheltered, low-lying area of dense growth containing a close network of trails where some of the trees have been browsed up to six feet or so from the ground by deer standing on their hind legs, you'll know you are in one and you should leave immediately because the deer need to be there and you don't. The wintering areas near me have changed a bit over the years, as the low, bushy trees of my childhood have grown tall and new young trees have sprung up in various patches, but at the time of this writing there are still a number of dense shelters. I never learned for certain who used them. One might ask why not, since I seem to be preoccupied with deer. I could have gone into the woods to find out, perhaps tracking the deer after they left the feeding area, which considering the snow would have been easy. The reason I chose not to is that I didn't want to disturb them, or scare them away from their shelters even for a short time. In a winter like that of 2007–2008, every calorie counts, and I felt that their need to survive far outweighed my curiosity or my desire for material to put in a book. Years ago, when I learned of the need of deer for winter shelters, I made sure that nobody went into our woods for any reason in winter. To cause a winter-challenged animal to spend precious calories in a needless escape is something even scientists should not do, let alone cross-country skiers and other recreation seekers, and certainly not writers, who if they can't learn about deer without distressing them should find something else to write about.

So I agree that the concerns of Fish and Game are very important, and here again, I think that having many different piles of corn may have been helpful. If a group of deer spends energy walking any distance for corn, meanwhile being exposed to wind and weather while away from the shelter, then the fawns must eat when they arrive. If not, they have depleted their energy reserves for nothing. Although big deer keep small deer away from food, having more than enough piles of food meant that sooner or later even the little ones usually ate something.

Did everyone get enough? I can't answer. Do I know what they ate? Indeed I do. I put the food in plain sight of my office, and could see not only who ate what, but also the size and number of the mouthfuls. I could count the grains of corn dust on their noses. The strong excluded the weak, as is deer custom. But everyone usually got something. Not one deer, not even the smallest of the Tau fawns, failed to eat eventually. Was what they got worth the trip and the exposure? Only the deer could answer.

Fish and Game points out that a deer expects to lose about 20 percent of its body weight from winter hunger even if food is present. This of course is well established for northern whitetails. According to various authorities, a deer can lose up to 25 percent of its weight, but if it loses much more, it's unlikely to survive. Nowhere does it say that deer need no food in winter. Deer need some food or they die. In one of very many similar experiments (this one conducted not just once but twice, the second to confirm the first), nine imprisoned deer who had no food except their feces starved to death, to nobody's surprise. Between their own meals, the well-nourished scientists would view their prisoners to see how the experiment was going. One deer died after sixteen days. Seven others died between twenty and thirty-seven days. The last deer died after forty-seven days.[1]

This kind of experiment, of which there were many, was supposed to determine how the intestinal bacteria of deer held up during starvation. Could the bacteria in a deer's digestive system withstand a long period of hunger? Would any bacteria still be there when the deer found food? In that particular experiment, even the bacteria didn't make it, or most of them didn't. The murderous experiment does point to one thing, however. Intestinal bacteria are necessary for digestion but aren't much use if there's no food, and this is something for deer-feeders to think about. By midwinter, deer have intestinal flora suited for winter fare, so corn or hay may be too rich for them, hay being different from dry standing grass because the hay is harvested while still rich in nourishment. The deer seem to have discovered this all by themselves, or so it seemed one long-ago winter when someone put a bale of hay in our field. The deer didn't touch it. It was still there in the spring. And on one occasion, also long ago, I put out special deer food, and the deer didn't touch that either. Perhaps their intestinal bacteria weren't ready for these foods.

However, in 2007 I began to distribute corn in the fall before I was fully aware of the lack of acorns. I meant it for the turkeys, and because I put it out at night to be ready for them before dawn, I didn't at first realize that deer were also eating it. They were, though, and at the time still had their nonwinter bacteria. It was a cause for concern nevertheless, so right from the start I checked everybody's droppings. The turkey droppings seemed normal enough, with what appeared to be the right proportions of brown feces, white urates, and liquid urine. How did I know about droppings? For many years I have shared my office with parrots whose health I monitor. Then too, for a time, our farm was a chicken farm. The only abnormality that I sometimes noted in the turkey droppings was a low amount of urates, which is said to be a sign of stress. And yes, winter was stressing them. As for the deer, they left scanty piles of round, winter droppings that also seemed normal, just what you'd see in the woods. In February, though, I found a somewhat soft, moist dropping. In it, I noticed a few apple seeds. Somewhere this deer had found an apple, and in consequence had left a summerlike dropping. And whenever

I'd notice a deer producing droppings, I'd mark the place in memory and would go for a look when all the deer were gone. This gave me an idea of how that particular deer was doing. And as far as I knew, all the deer who left droppings were doing fairly well.

It's impossible to speak of wildlife without speaking of droppings, more often called scats, because a scat is like a little book packed with interesting stories. Often a scat has a point on one end, for instance. This was the inner end, the last part of the scat to leave the animal, and thus points in the direction the animal was going. A scat that looks like a dog's but contains hair and bone fragments is probably a coyote's, and if the hairs are about two inches long and fairly stiff, you can feel pretty sure that this coyote has found a deer carcass. If you find a group of scats that look like black rice below an outdoor shutter, a little brown bat, probably a male, is spending the daytime hanging behind the shutter. A thin, hairy, somewhat sculptured scat, two to five inches long, up on a boulder, may be a bobcat's who has purposely placed the scat where the scent will float out to discourage other bobcats from encroaching. Any scat up on a boulder, bobcat or otherwise, is someone's KEEP OUT sign. If you come upon a scat that is rather nondescript like a person's, about an inch and a half wide and four inches long, perhaps with one or two smaller portions detached from the main body, such as a person might produce, it's probably a black bear's. If you open it up you may find such things as hair or berries or beetle carapaces, which will show you what the bear has been eating. If you find a scat shaped like a hot dog about five or six inches long and somewhat pointed at both ends with hair sticking out, parallel to the body of the scat, and if white bone fragments show through the sides, it could be a mountain lion's and the skin at the nape of your neck should prickle a little. Keep watch behind yourself as you move off.

As for deer scats, for most of the year they are small, dark, and compact—typical of many animals such as porcupines and rabbits who eat bark and other rough vegetation. Porcupine scats are oblong. Rabbit scats are round but somewhat flattened. Deer scats can be round or oval, sometimes with a little point on one end. In winter they emerge from the deer individually. Sometimes you will see them in large piles of fifty or more pellets, and sometimes in small piles or scatters. In summer when the deer are eating more lush foods, the droppings tend to be somewhat softer. This causes the little scat-balls to flatten out and stick together, hence they may appear as a single, layered scat, like a cow's but smaller and more defined. If you see a loose, damp-looking deer scat in winter, you can be fairly sure that somewhere out there in the woods, a deer is sick. With the exception of the scat with apple seeds in it—not a sign of illness—I found no scats like that.

All the deer ate corn sparingly, but more so in midwinter. I put corn out every day with no exceptions, always early in the morning, and only twice did they finish all of it. Later in the year they began to eat more. I put out more. By March, they again were eating more, and I put out even more—fifty to seventy-five pounds a day or about two pounds per deer at their maximum number, although, of course, fifty to seventy-five turkeys were sharing it. Meanwhile all the deer dug under the snow for dead leaves and grass. They also ate the twigs of trees and bushes, and rotten wood from a dead stump near the house. These are foods they would have eaten anyway, because these were their natural foods, however low in nourishment. By the middle of March, a whole day could pass with only four or five deer coming. I took this to mean that they ate the corn not because they liked it but because they needed it, and they turned to other food such as frozen dry grass because they liked it better.

Late March is a hard time for deer, because their metabolism is returning to its nonwinter mode. But even in an ordinary year, the foods in the woods don't necessarily supply them. Toward the end of March 2008, the snow in the woods was still deep, the twigs seemed as lifeless as they had all winter, and the deer were coming in large numbers again, seldom less than fifteen together, twenty-five on one occasion. Evidently, their metabolisms were readjusting. They seemed truly hungry, and they no longer worried about Group Alpha. By then, some of the snow had melted in the field, and patches of dry grass were showing. After eating corn for a few minutes, the deer would go to these open places and eat grass, sixteen or seventeen of them together. Then they would find a sunny spot, perhaps on the island of trees, perhaps in the south-facing microclimate under the oaks at the edge of the field, where they would lie down to chew.

Like all ruminants, deer have divided stomachs to deal with their food. Thus they can eat for as long as an opportunity lasts, packing the food into the first division. When they feel safe and have time, they lie down, bring up the food in little balls about the size of plums, chew each ball thoroughly, and swallow it into the second division. This must be a very nice experience for deer—a sunny spot, a chance to rest, and the pleasure of chewing and swallowing. Nobody knows which division, when full, gives a deer a sense of satisfaction. Perhaps both, first one, later the other. Or perhaps the two divisions give different messages when full. Perhaps a full first division gives a sense of security, while the second when full gives comfort and pleasure.

On March 26, to my very great joy, I noticed one little Tau fawn lying in the sun all by herself, calmly chewing. All around her, other deer were grazing. Among them were her mother and one of the grown daughters. I was elated that these three, at least, were still alive, and that the fawn had already eaten enough that day to require a long session of chewing. All present seemed relaxed and comfortable. They were as far from the house as they were from the edge of the woods, out in the open where nothing could surprise them. They didn't need to keep watch carefully. So those who were grazing looked up rarely, if at all, and instead kept their heads down, eating, eating. Then the sun set and the temperature plunged. The little fawn must have felt chilly—that, or she had finished chewing. She stood up and went to graze with the others. Just before dark, all these deer came back to the feeding area, twelve of them including the Tau fawn, and all ate corn together.

Fish and Game points out that the spread of disease among the deer herd is an important peril. As of 2007

the terrible chronic wasting disease of deer had not been found in New Hampshire, but it was in New York State, threatening to enter Vermont and New Hampshire at any time, and would be a very important danger. Deer suffer from other diseases too, so in an effort to minimize the risk of a sick deer infecting others, I adopted a clean plate system, each day putting the piles of corn in new places on fresh snow. Here, the frequent storms were helpful, every few days providing more fresh snow. Because there were more piles than deer, the deer tended to eat individually, and those who ate from the same pile were already in contact—perhaps a doe and her fawns. The clean plate system also protected the flock of turkeys, who ate there daily, sometimes with the deer. It did not, however, protect deer from coming near one another, or from eating from a pile where an infected deer might have been eating earlier. So the clean plate system might have helped, but it didn't protect altogether. However, I felt very sure that without the corn, some of these deer would not have lived through the winter, so I took the chance and was glad to see in early spring that the deer were still healthy. This only meant that we were lucky, and that thanks to luck, illness had not been a problem.

A large aggregate of deer can attract predators, says Fish and Game. This could be true of some areas, perhaps, but not ours—we already had predators. One summer, a friend saw a lion in our swamp—an African lion, he said. Could global warming be worse than we thought? Or could he have seen a mountain lion, possibly the same one I saw near that very swamp a few years earlier and whom other people also reported? If so, deer-feeding hadn't lured him, although the deer themselves might have. But if he was still in our area at all, he came before I began to put out corn.

We also had black bears, but they were asleep. We also had coyotes, a pack of whom held the coyote land rights in our woods. They preyed upon the deer just as they had always done, but they didn't do it easily and they certainly didn't want competition from other coyotes. If other coyotes had tried to move in, they would have been firmly discouraged, and if they had succeeded, they would have evicted the original owners. No matter what the outcome, the deer would have been dealing with just one pack of coyotes.

Interestingly, it was the turkeys who seemed to deal most effectively with coyotes and made an important distinction between them and the dogs. If the turkeys saw a dog in the daytime they would merely trudge off in the opposite direction, but if they saw a coyote, such as the spectacular, large-size, heavily furred coyote who one day came walking along just inside the edge of the woods, all fifty-five of the turkeys would form a pear-shaped group with the boldest turkey in the lead, and they would follow him. Surely that particular coyote had an interest in them, but he pretended that he neither knew nor cared that they were close behind, and he went calmly onward. What, after all, is the use of spooking potential prey animals who already know where you are and are watching you? The more innocuous you seem, the more likely they are to forget about you. So, showing with his relaxed ears and lowered tail that he wasn't thinking about the turkeys, he trotted along, blameless as a newborn lamb, keeping just inside the woods, with the turkeys keeping pace with him in the field. When he reached the end of the field and was heading into deeper woods, the turkeys stood still for a minute, perhaps wondering if their task was completed, then seemed to conclude that he posed no further danger, and they dissolved their pear-shaped flock and went on about their business.

The coyote did well to stay away from them. Turkeys are extremely strong and are excellent fighters, and they outnumbered the coyote fifty-five to one. Turkeys have been known to attack and seriously injure grown men, and once, down by the road, one of them attacked our dog (a dog who was no longer living when I began to feed deer), as we learned when we heard her screaming as if she'd been hit by a car. We ran to her aid and found her cowering under some bushes. In a nearby tree a big tom turkey was looking down at her with his feathers all fluffed.

Whether or not the deer attracted coyotes, as Fish and Game said they would, the turkeys did, and therefore the coyotes would have been there anyway. Before the turkeys were enjoying our corn, we heard coyotes almost every night. After that, we rarely heard them, and only far away. But they were not always far away— at night they came in silence, leaving tracks in the fresh snow that we would notice in the morning. The turkeys wanted to be ready for the corn and would roost in trees at the edge of the field. Before sunrise they would glide or soar down to the ground and form their groups, getting ready to walk to the feeding area together. If during the night the coyotes had been flaunting their presence, the turkeys would probably not have glided to the ground near the edge of the woods, but would have spent energy to fly farther out in the field. As it was, the flock of fifty-five turkeys in December was reduced to fifty-one by February, and during that time, just behind the tree line, two silent coyotes would pass by to check on them every few days.

Our area also had a bobcat who lived on the slopes of North Pack, the mountain east of our house. Bobcats can sometimes manage to kill small, winter-weakened deer, but we did not expect to see more than one bobcat, no matter how many deer were present. In New Hampshire, a bobcat controls at least five square miles as his personal range, and he tolerates no intruders. Nor do others attempt to intrude, due to the rules that bobcats live by. Even during mating season, bobcats don't live together. The isolationist tendencies of bobcats became very clear after a friend, Sue Morse, and I once put a squirt of mountain lion urine on a tree in the woods. How did we come by mountain lion urine? When I saw the mountain lion near our swamp I wanted to lure it back, so I called Sue, who asked another friend, a zookeeper in Arizona, to send us by FedEx some urine from a mountain lion in his care. (Sue is the kind of person who can arrange such things, to her eternal credit.) The urine we sprinkled did not lure the mountain lion but was noted by the bobcat, to his horror. He sprayed, scratched, and defecated, wildly and furiously, sending a forceful message to whoever left the urine that the area was owned, and trespassers would be dealt with harshly. It seemed safe to say that the deer of our area were coping with just one bobcat.

The very fact that the feeding area was near a house also offered protection. One morning when deer and turkeys were eating, the bobcat walked out of the woods beside the pond and stood still in plain sight, watching them. He was several hundred feet away but even so was very conspicuous against the white snow, and the deer and turkeys noticed. All raised their heads and looked in his direction. Normally turkeys make a very distinctive, questioning sound if they see a predator, and deer either leave or stare and stamp, but not on this occasion. They simply watched him. And he watched them. He must have been hungry—he watched them in the same way the low-ranking deer sometimes watched their social betters eating corn, wanting some but doubting they would get any. Nothing about the bobcat showed that he was hunting—he was just looking. My guess is that as far as he was concerned, there was just too much going on up near our house, with the dogs and the cars and the smoke from our chimney, to say nothing of the house itself, with reflected sunlight blazing from its windows. At last he turned away and went into the bushes.

One day in late March a doe went by with long claw marks on her left shoulder. These showed conspicuously in her fur, but were not bleeding. Whoever had grabbed her had big hands with claws about an inch apart, and had set his claws near her neck and dragged them down and back toward her flank. Since she wasn't attacked near our house or feeding area, I could only guess what had happened. The marks were in the exact place where a big cat would have grabbed her. Had a cat of some kind attacked? The mountain lion, maybe? A mountain lion who got that far with his hunt would most likely have succeeded. He'd be big enough to knock her over, and that would be that. The marks seemed much too wide to be a bobcat's, and bobcats rarely attack full-size deer. So my guess is that a black bear made a grab at her—a hungry bear just out of hibernation. Bears are fast, if not as fast as mountain lions, but deer are the fastest of all. Evidently, the young doe made a successful dash for freedom and escaped. Could my corn have given her the strength to do so?

Speaking of predators, an additional peril to deer fed by people is that they may become habituated to our species and fall easy victim to hunters. But the deer I fed did no such thing, and with the exception of a few brief events, such as my roadside meeting with the Tau Group, the deer continued to fear us.

I may care about deer, but I don't expect them to care about me, therefore we had no trace of a relationship that winter and made no effort to forge one. If I happened to leave the house when deer were approaching, they would stand and watch me. But the moment I disappeared around a corner, they would bolt, even if I had been moving away from them when I vanished. As long as deer can see the predator, they feel reasonably secure if she's far away, but when she's out of sight and they don't know what she's doing, they run away as fast as they can, believing that the predator's casual manner was merely a ruse to put them off guard. This is the classic reaction of many animals to faraway predators, who seem safe enough when they are in view and appear not to be hunting. But who knows what they're up to when they are out of sight? The deer of our area were afraid of me at all times, and of all people at all times, and hopefully they always will be.

Finally, in the list of reasons not to feed deer, Fish and Game mentions the damage they do to ornamental shrubs. This is certainly a problem to some, but as far as I'm concerned, they can eat my shrubs down to the stumps if they feel like it. In my cosmology, indigenous wild deer are more important than exotic ornamental shrubs.

All in all, I'm not convinced that my deer-feeding did more harm than good, at least in the winter of 2007–2008. My favorite authority on deer, Dr. Rue, has this to say about starving deer: "When you hear preservationists talk about 'letting nature take care of its own,' you know they have never witnessed deer dying of starvation. They have not seen a beautiful deer turn into a dull-eyed, listless rack of bones, wrapped in a rough coat, stoically awaiting death. The deer don't comprehend what is happening to them. There is nothing they can do about what is happening, so they only attempt to endure."[2]

Dr. Rue is strongly opposed to state-controlled programs of deer-feeding but adds that "feeding deer on an individual basis or in special situations is not wrong."[3] He cites a personal friend who for many years fed cracked corn to more than fifty deer in winter, always with good results. The fat that deer accumulate throughout the summer and fall is their lifeline through the winter, but they can't survive if their fat gets used up. So if the corn I fed prolonged the fat reserves of the deer who ate it, I probably struck a balance. And considering that a new storm blew in every few days, bringing heavy snow in the middle of winter and snow mixed with heavy, freezing rain in March—the kind of storm that wets animals to the skin and makes it very hard for them to keep their bodies at a viable temperature—they probably didn't suffer as much as they would have without help.

I have the greatest respect for New Hampshire Fish and Game and the University of New Hampshire Forestry and Wildlife Program, and I understand the protection that they wish to offer. Over the years, I have put into practice many of the conservation measures that they recommend—the encouragement of nut-producing trees, a managed forest that includes two substantial clear-cuts where tender young trees and berries grow, long grass in the fields where deer can sleep at night from June until October, and no development. I wouldn't even have fed the deer if the oaks had not been self-protective.

Nobody wants disapproval, least of all me, so I hope that some of the measures I took to safeguard the deer from my well-intended efforts might mitigate some of the disapproval that others might feel. But probably not. Plenty of people feed deer, it's true, but they do so fairly surreptitiously and they don't write how-to books about it, so I expect plenty of criticism from some who may learn of this book, not only from conservationists and other wildlife experts but also from home owners whose shrubs are chewed, from drivers who wreck their cars by crashing into deer, from people who fear Lyme disease, although it's ticks, not deer, that cause it, and from macho types who find deer-feeding sentimental and mockingly refer to Bambi. All this is saddening, of course.

Well, it is and it isn't. Deer do not oppose deer-feeding.

Chapter Five

Deer Seasons, Human Seasons

What is life like for a deer? Deer are indigenous to the New World forests, and the deer who live in the North are equipped for dramatic seasonal changes. We were indigenous to the African savannah and are equipped to survive in the sun, which is why we're as naked as mole rats, except for the insulating hair on our heads that once kept our brains from baking.[1] Thus an interesting way to approach the differences between ourselves and the animals whose northern world we were later to invade is to consider our reactions to the sun, particularly those that occur without our knowledge.

Like most vertebrates, we have tiny holes in the backs of our eyes through which light shines upon our pineal gland as it measures such things as the amount of sunlight and the hours of daylight and dark. If we don't see enough light we get depressed, perhaps because, due to our ancestral memory of the savannah, too much dark suggests that we are very far from home, or worse yet, that an important celestial malfunction is in progress. Who wouldn't find these messages depressing? We carry the memory of reliable bright skies where days and nights were more or less the same length and every season was a hot season, at least in the daytime. Our message from the sun was to stay in the shade.

But suppose we had evolved in the northern forests, rather than simply arriving there as an invasive species. We certainly wouldn't be naked—we'd be permanently covered with dense fur—and when our pineal glands told us that the days were getting short, we'd do a lot more than simply feel gloomy—we'd redouble our efforts to find food, and we'd start breeding so that nine months later our young would be born in the spring. Allegedly we do eat and breed a bit more in the autumn, but if we were a truly northern hemispheric species, we'd do it in grand style. And all around us, our fellow northern hemispheric animals would also be responding to pineal information—they'd be growing extra fur or migrating or trying to get fat. These are seasonal adjustments, but we evolutionary strangers are poorly equipped to appreciate them. The reason we don't have thick fur and a breeding season is not because we're superior beings, but because we evolved where such things were not needed. Most of us are so taken with ourselves as a superior species that we don't like to think along these lines, but it's useful to try.

Whitetails, in contrast, are tied tightly to the seasons. Much of the literature about deer tends to treat the seasons equally, as if the deer did too, but it became my impression that as far as the northern deer are concerned, there are just two seasons—winter and nonwinter. In winter, deer cling to life as best they can, trying to escape death from starvation and hypothermia. In nonwinter they do everything else—give birth, eat, grow, mate—all of which are timed in order to best survive the next winter, and are governed by the sun. When the hours of daylight seem sufficient—during the first week of April in my area—the pineal glands of male deer send hormones to start their antlers growing. Round lumps appear on their foreheads about halfway between the eyes and ears, but a bit closer to the eyes, and about thirty millimeters in diameter, more or less, depending on the owner's maturity. The hair over these buds seems light-colored because, as with many animals, each hair is lighter near the skin than at the tip and the stretching skin separates the hairs and lays them open. No longer is it hard to tell males from females.

In April 2008, I could reaffirm my guesswork as to who was who, because even the male fawns, who by then were almost yearlings, also had antler buds.

It takes time to grow a pair of antlers, but by mating season they must be ready. An antler grows about a quarter inch a day inside a soft coating filled with arterial blood, a nourishing tissue known as velvet. Antlers are bone—the calcium to make them comes from the deer's own body—but at first they are soft and easily damaged. Deer tend to hide while their antlers are growing. From May on through the summer, I saw male deer only twice, both times in the evening, both times at the edge of the woods. I was touched to see that they came in groups, several youngsters with a big, important leader.

By September, antler growth is complete, and the velvet that nourished the bone falls off in strips. By then the bone is strong and the deer resume normal activity, feeding themselves as best they can to prepare themselves, first for the rut or mating season, later for winter. Here again, winter looms, and timing is everything, because fawns obviously can't be born in winter. Nor should they be born too late in the summer, and certainly not in the fall, because they won't have time to grow big enough to survive the winter. Thus spring is the best time to be born. The gestation period for whitetails is six and a half months, which means that in New Hampshire the rut must start in November.

And so it does. Deer are infertile for most of the year, as are many other animals who have a mating season. Year-round reproduction is fine for those who don't have to time the birth of their offspring, but not for those who do. We as African savannah animals could afford to keep our sexuality year-round, but the deer cannot. Thus the rut lasts a fairly short time, and the deer must make the most of it.

As the rut begins, the testicles of the males drop into the scrotum and live sperm begins to form. And the females start to cycle. Just like us, every twenty-eight days (some say in response to the moon), they experience a one-day period of fertility. Males must locate them and protect them from the advances of other males until the fertile day arrives. This can involve some fighting, most of which is mild.

On many occasions over the years, we have seen bucks sparring in our fields, usually with two or three others watching, as the sound of clashing antlers draws other males as spectators. Why do they spar in the fields, especially in hunting season, when they could become easy victims? Of course, they are distracted by the rut, and are focused on their rivals, but also, I believe, because sparring requires a certain amount of space. The moose of our area, who are seldom seen because they stay perpetually in the woods in the daytime, need even more space. Sometimes they also come into the field to spar with one another.

Much of deer literature holds that deer fighting is serious and potentially lethal. Some of the descriptions are quite frightening. Sporting magazines sometimes have drawings of bucks that look mean and angry, with narrowed eyes and downturned lips, as a person would look if he were just waiting to get his hands on someone he didn't like. This might make hunters feel good about killing deer, but it isn't quite true. I've seen many deer fights. None were serious. Interestingly, a large male will sometimes invite a young male to spar, and the young male will do so with no hope of winning. It's more like a game. Throughout the year, young males associate with older males who teach them virtually everything they know, so surely these sparring sessions are part of the youngsters' education. Then too, the older male wants to make sure that the youngsters know his power, as they will be tagging after him while he searches for females, and he doesn't want trouble. As long as they understand their status, their presence might help by discouraging other males from approaching a daughter while he impregnates the mother. At such a time, a young male may even get to service the daughter while the big guy is busy. Try as he may, he cannot impregnate them all.

Typically when equals spar, the combatants eyeball each other for a while, then, from a short distance, charge together but not too fast, then push each other until one of them decides he's had enough and backs off, temporarily accepting the fact that perhaps he's the lesser deer. That settled, all of them then stand around for a while, as if wondering what to do next, and all eventually move off, sometimes together and sometimes in different directions. This might be called fighting, but it is very far from serious fighting.

I have never seen a truly serious fight, and the only circumstances I can imagine that would cause one would be if one very dominant buck appeared in the territory of another. Then the two might clash more dangerously, and this could cause accidents. The antlers of two bucks can lock so that they can't untangle. They are locked together for good. Sooner or later, something will probably come along and kill them, and if not, they will die of starvation. Another possible result of a serious fight is that one buck might hook the other in the body. Antlers are sharp and can make a deep wound. Deer have enough problems as it is, and more problems if they are wounded. Hence the less aggressive sparring.

One of many factors that may mitigate aggression is that the deer of an area are usually related or acquainted. Even if they are not together, they inform one another by means of their scrapes—shallow depressions on the forest floor, sometimes round, sometimes oval, that can cover two or more square feet. The males make these scrapes, which involves an interesting, complex process. If you know what to look for, you can see the scrapes, as they are intentionally conspicuous. They will be on a deer trail in a relatively open area, usually under a tree that has at least a few branches five to eight feet off the ground. Why? Because the buck wants his scrape to be obvious, and to this end he stands on his hind legs and breaks one of the branches so that it hangs down. He may also chew the branch a bit, and rub his forehead on it, which imbues it with his scent. He then makes his scrape in the place to which the branch is pointing. But not every scrape has a broken branch. To the west of our house is a forest of enormous old trees with branches high above the ground, and deer make scrapes there too, even though they can't reach the branches. These scrapes seem wider and deeper, as if to compensate for the lack of pointing branches. In other words, a deer will make a scrape because he's in rut, and will make it on a deer trail, and if he can't find a proper tree he'll make it anyway, but he greatly prefers that a branch with his scent on it points to it.

It once was said that the ability to use tools was unique to human beings. Then it was discovered that other primates also use tools. Then it was discovered that birds use tools. In short, many kinds of animals use tools, and in my view, deer do too. If creating a marker that holds your scent is not using a tool, then I don't know what is.

The scrape itself might be considered a tool. The buck makes it by clearing away the fallen leaves and some of the earth with his front feet. When it is of a size that's to his liking he stands on it and rubs his rear feet together to collect the pheromones from the glands on his ankles. He then shifts his weight to his front legs and brings his hind legs forward so he can urinate on these glands. I've never seen a buck mark a scrape, but others have, and evidently, for the urinating part, the buck assumes a unique and astonishing posture—some say it's almost a handstand. I'm sure this happens, but the only times I've seen a buck urinate on his feet, he squatted like a doe, although his hind feet were farther forward than a doe's would be. At any rate, he washes a meaningful mixture into the scrape, which tells the other deer that it was he who made it.

A scrape also becomes a signpost, and this, to me, is why it's like a tool. The buck who made it will refresh it from time to time, but other passing males will also mark it, just as dogs and other animals overmark the urine stains of their contemporaries. The deer of an area may know one another, but they don't necessarily stay close together. Usually their little groups are spaced quite widely. A scrape thus serves the entire community, as passing deer will know not only which others are nearby, but also will know who is rutting.

Every autumn by our swamp, the deer make scrapes on the trail often used by the Deltas. I go there now and then to see how the deer are doing. In 2006, I found just one scrape, very large, with leaves and earth cleared in an oval of several square feet. A large branch pointed to it. Whoever broke the branch was big—the break was almost eight feet off the ground. But in 2008 when I went for a look I found three much smaller scrapes and felt quite sure that the important buck of 2006 was not the one who made them.

This spoke of a tragedy. A neighbor told us that in the fall of 2007 an enormous ten-point buck was shot but not killed by a hunter who was hunting illegally on our neighbor's property. The land was posted and the hunter did not have our neighbor's permission. The buck died of his wounds, and his corpse was found by the game warden. Whoever shot him must not have been much of a tracker, nor much of a hunter. Rather, he was an incompetent, irresponsible poacher who stupidly destroyed the most important member of the deer community. That buck was surely the father of most of the deer I had been watching, certainly the younger ones. Unquestionably, it was he who had made the big scrape. Lesser bucks must have respected it and added to it, keeping it refreshed and clear. If they also made their own scrapes, they did so somewhere else.

The three new scrapes confirmed his absence. One seemed dysfunctional. The branch intended as a marker was thin and brittle and had broken off completely. It lay beside the scrape, which had almost filled up again with leaves. Perhaps a small, unimportant deer had made it. On either side were two other scrapes, not very big but relatively clear of leaves, both with branches pointing down at them, but here again, the branches were not much better than large twigs, and the breaks were only about five or six feet above the ground. Smaller bucks had broken these branches. To be sure, that place had been used for scrapes for many years, so suitable branches were no longer plentiful, and the governing factor of where to make a scrape would seem to be the presence of the deer trail more than the quality of the branches. But clearly the scrapes of 2008 were not as big or as well made or as often visited as was the scrape of 2006.

The death of the ten-point buck diminished everything and everybody. On November 30 I saw a certain yearling with spike antlers out in the field in the daytime, grazing. A tresspassing hunter would have shot him. If the important, knowledgeable, ten-point male had been alive, the yearling (who might have been his son) would have spent time with him and would have learned from him that deer, especially antlered bucks, stay out of sight in the fall in the daytime. I scared the youngster away, but he came back. Deer can learn about people, but not from them.

During the rut, a male will try to impregnate as many females as he can. Some, such as the murdered ten-point buck, will impregnate most of the females while others may impregnate none. But the females may or may not keep the semen. If a doe wants to retain the semen, she will stand still after copulating, with her back arched. She also may squat and strain as if to expel it. Why would she do this? Many animals have preferences, so perhaps she would rather carry someone else's fawn. Or perhaps she chooses not to be pregnant at all. Yearling females cycle but don't often bear fawns, and Dr. Rue has seen them squat and strain for as long as half an hour after being bred, as if to thoroughly expel something. He himself didn't see a discharge, but he cites others who did—a "yellowish secretion."[2] Was it semen? If so, would a young doe expel it because she isn't physically ready for pregnancy? Or might she be wary of her social betters, the mature, high-ranking females, who don't want her fawn competing with theirs? Some social animals such as wolves can be very careful not to get pregnant while in the domain of their social superiors, and deer are no less hierarchical.

But how must this affect a male? After all his scraping and rubbing, after all his sparring and fighting, what is he supposed to do if the hard-won doe he has just bred squats and squeezes out his semen? Should he keep tending her in hopes that he can breed her again? Should he go looking for a more willing female? If she gets rid of his semen she will cycle again, so if he leaves her, will he be handing her over to one of his rivals?

I saw a buck in such a dilemma in our field in mid-November 2008, the year after the ten-point buck was murdered. That day, a four-point buck tried to mate with the younger daughter of the Deltas. This was an unusual sight, as hunting season was in progress. I knew for sure that people were hunting illegally, either on my land, which is posted, or on the Wapack Wildlife Reserve—also posted—on North Pack Mountain, which abuts my land to the east. Now and then we'd hear a gunshot. Normally, deer seldom appear in my field in the daytime during hunting season, except to spar occasionally. These deer did, though. The Delta daughter was unworried, and the buck couldn't resist the scent of estrus. He spent quite a while sniffing under her tail or trying to. She was grazing and wasn't very interested in him, although she didn't rebuff him. At last he mounted her, extruded his pale penis, and penetrated her. She stood still for a few moments, but then pushed backward and dislodged him. She went on grazing, making her way toward the woods. He followed at a distance. It wasn't clear to me whether or not he had ejaculated. Perhaps not. But even if he did, he could not be sure he had impregnated her and would still need to follow her to try to keep another male from mating with her too. Only if she stopped cycling could he be sure.

Obviously a male in his situation can't just relax and start eating for winter. He must keep on scraping and rubbing, threatening the smaller males, cautious of the bigger males, sparring with his equals. Testosterone is flaming inside him, burning up whatever fat reserves he may have gained in summer, and only if he finds sufficient food in early winter will he be able to restore himself to prime condition. This doesn't always happen. Therefore male whitetails are often vulnerable to winter, dying at a rate that requires the species to compensate. More males are born than females. We do that too. Baby boys outnumber baby girls, allegedly to compensate for our warlike inclinations. Like war, a rut is conducted at the expense of life and profoundly engages all those who are involved with it, at enormous cost to one but not both genders. As far as male deer are concerned, the rut is the ultimate distraction.

Our response is to open the hunting season. Taking full advantage of the needs and distraction of the deer, we fill the woods with invasive primates camouflaged to look like piles of leaves who sneak around, sprinkling estrus doe urine and manipulating gadgets that sound like antlers clashing, all designed to trick rut-consumed males into coming in their direction. The obvious quarry of most hunters is the large bucks, those with impressive antlers who have managed to survive several winters and who, if they were left alive to breed, would pass their abilities to the next generation. It goes without saying that a deer population is better off if the strongest, most vital animals are alive and breeding than if their heads are on somebody's wall. However, a certain amount of semi-science is offered to support their slaughter—when they're gone, it is argued, the younger bucks can breed and thus diversify the gene pool. In reality, of course, the little bucks are probably the sons of the big eight-pointers and won't be offering much diversity. It's complicated, of course, and a number of truly scientific studies have been done on exactly this issue, but the long and the short of it is that any population that manages itself is following the evolutionary path set by Gaia, and is normally better off than if another, less knowledgeable species such as ours interferes with it.

During the rut of 2007, the deer were doubly disadvantaged by seemingly unlimited snow in which they had no choice but to leave footprints, enabling hunters to track them. The snow also dampened and buried the dry autumn leaves, which in a normal year would make rustling noises when stepped on and would warn the deer of an approaching hunter. In the snowy autumn of 2007, the hunters walked in silence. That year in New Hampshire, the fatalities were legendary—an estimated 13,416 deer were killed, or 15 percent of New Hampshire's deer population. A similar slaughter in human terms would eliminate all the people in two of New Hampshire's biggest cities, Manchester and Nashua, along with the populations of several towns. The number of reported fatalities does not include deer who were shot but escaped the hunter and died later. There are always some of these too—witness the corpse of the ten-point buck on our neighbor's property—but no one knows how many. At least the corpses feed the bears and bobcats and coyotes who also will be trying to survive the winter. With all due respect to the nation's Fish and Game departments, more deer die because people hunt them than because people feed them. No matter how opposed to deer-feeding some people may be, no one has ever suggested that extra food in winter could result in a death toll of such magnitude.

The rut tapers off in December, just in time for the most serious part of winter. In response to the fading rut as well as to pineal information, the testosterone level in the males drops, so the males cast their antlers, first one, later the other. After midwinter comes the equinox, a signal that nonwinter is at hand. The balancing hours of daylight and dark inform whitetail metabolisms to get moving. All the deer who came for my corn seemed hungry, but very little food was in the woods. However, twigs were budding and the days were getting warmer. At the edge of the pond, the beavers began to cut down a tree that would fall uphill, butt end toward the water. On April 17, two nights before the full moon, frogs began to sing in the swamp.

At about that time, the deer resumed their nonwinter behavior. For the past six weeks or so, I had watched my deer arrive less regularly, without the Alpha Group, who might not have wanted to cross the weakening ice, and without the Taus, who may have been unable to come and may have suffered terribly. None of the deer came every day, and at last only two groups came reasonably regularly. These were the Delta mother and her two daughters, and the five wonderful Betas—the mother, her grown daughter, and her three fawns, including the fawn she may have adopted. In the late afternoon, I would see these deer in the field with their heads down, grazing. During a period of about two weeks, they seemed to leave the woods exactly when the shadow of the hill touched the edge of the woods and the sun stood four fingers above the horizon. (For those unfamiliar with the ways of the woods, an ancient way of telling time is to hold one's hand at arm's length so that one's fingers are between the sun and the horizon.) The deer timed their arrival precisely. Day after day I'd watch for the shadow and note the elevation of the sun. And just as the deer started up the hill, the shadow would touch the edge of the woods and the sun would be exactly at four fingers. What had they been waiting for? The shadow? The elevation of the sun? Something else I hadn't noticed? I couldn't answer my own questions because the deer behaved in this manner only for a short time, then gave up the interesting practice. They continued to come, but later.

When the moon rose, the two groups of deer would go back to the woods, the Betas to the ridge to the east, the Deltas to the trail past the swamp. Later in the night both groups would reemerge to make their way up the hill and lie down in our field in the moonlight. If I came up the driveway at night, the deer would stand up and my headlights would shine in their eyes, first into five pairs of eyes at various distances from the ground on the right side of the hilltop—the Betas—and then, as the headlights swept around to the left side, into three more pairs of eyes, all about the same distance from the ground—the Deltas. Right through the end of May I'd see their eyes shining, but when the weather grew hot I didn't. Perhaps the deer were deep in the woods, keeping cool, doing their best to avoid the black flies and mosquitoes. Or perhaps I simply couldn't see them because the grass was too long. I know they sometimes came, though—a sleeping deer curls up in the grass, which flattens it into a deer-size circle, and I'd find their empty beds.

With the winter of 2008–2009 just seven months away, the next task for deer was to give birth. At this time, the mature, pregnant females supposedly chase off their daughters and live alone, or they do according to the literature. The various authors seem quite specific. For instance, according to one of them, "The whitetail doe isolates herself on a small territory, driving off other deer, including any of last year's young who are still tagging along."[3] I do believe this to be partly true, not just because it pervades the literature, but also because, one spring, I happened to see a doe trying to chase away her yearling daughter. This took place on an old logging road in the woods to the west of our house. I was surprised. Normally, the only other times I saw deer in the woods, I was up in the tree by our swamp, hoping to see wildlife. Now and then I'd see a coyote or a fox, and now and then a beaver or two swimming down their homemade channels in the swamp water, but only rarely would I see a deer go by. Therefore, just to be walking through the woods and chance upon a deer seemed unusual. Yet there they were, the two of them, about fifty feet away, going in the same direction that I was.

They didn't notice me. That also was unusual. They were preoccupied with each other. The daughter was trying to nurse from the mother, but the mother kept driving her away. The daughter would try to follow the mother, and the mother would turn on her and kick her. There was no resolution to this scene—I stood still while the two crossed the path in front of me and while the mother jumped over a stone wall into a thick part of the woods, as if running away from the daughter. Despondent, the daughter followed slowly. The two reemerged farther down the path, the daughter still begging, still trying to nurse from time to time, and the mother still dashing away or kicking at her. Many animals—dogs, for example—are good at judging whether or not their efforts are productive. If not, dogs stop trying, thus saving time and energy, which is to any animal's advantage. Dogs are very good at this, belying the word "dogged" as a synonym for "tenacious." I don't know how long a deer will persist, but it's longer than I would have thought, because when I saw the two for the last time, perhaps half an hour later, the daughter was still trying to nurse from her mother.

What did this mean? My guess—having watched deer more carefully since that time, and considering the season, which was spring—is that the mother had recently given birth. Her newborn fawn, of course, would have been hiding somewhere, but the milk in her udder would have been giving off its scent, and the yearling could have taken the scent to be an invitation. Maybe she thought the milk was for her.

Surely she remembered nursing. Most animals, certainly the so-called higher vertebrates, have enviable long-term memories. The yearling probably remembered the satisfaction of milk, her mother's warm flank, the summer sunshine, the security. Since then, she had survived her first winter. Young as she was, she had learned as much as anyone can about hunger, the freezing wind, the heavy snow—and during that time she might have thought about the meals of milk she had enjoyed a few months earlier. Perhaps her mother had nursed her through at least part of the winter, as a doe will sometimes do if she has milk. The practice may have saved many a fawn from starving. Perhaps the yearling was still thinking about this in the spring. The winter must have been hard on her. She seemed very needy.

Thus in the spring of 2008, I found it interesting to see the Betas and the Deltas still together. This alone cast some doubt upon the massive literature about whitetail birth, and my guess is that at least some of the various authors got their information from reading one another's publications. As Dr. Rue points out, very few people have seen a doe give birth, alone or otherwise. He has seen several births on deer farms, not exactly wild conditions, but similar. On one occasion he was able to witness a primapara doe in labor that can only have been extremely difficult, but probably was not abnormal. The doe was alone when her water broke at nine o'clock in the morning, and the fawn wasn't born until one o'clock in the afternoon after a hard and very painful labor. Most members of the deer and cattle families give birth standing up, at least for some of the time if not for the entire process, and so do whitetails who have given birth before. This seems to be as true for other mammals as it is for people—any woman can tell you that the second or third delivery is easier than the first. In one case, according to an observer, the doe went right on walking and browsing as the fawn slowly slid out of her. She seemed to scarcely notice, let alone experience discomfort. But some whitetails lie down, getting up from time to time, just as many women want to under similar conditions. A standing cow or hind arches her back and tucks her hips to let the weight of the calf pull it down and out. But the doe that Dr. Rue was watching lay twisted on her side, moaning bitterly and bearing down hard, with only the fawn's front legs emerging. When the laboring doe struggled to her feet, the fawn's legs slid back in. Dr. Rue tells us that she moved around restlessly, moaning and suffering, coming back to her chosen spot to lie down and try again. After four long hours, she delivered her fawn. Dr. Rue wrote, "It was the hardest birthing I have witnessed."[4]

In the wild, a doe in these circumstances is extremely vulnerable, especially if she is having more than one fawn. There they all are—the mother consumed by pain, one fawn still in her birth canal, another fawn weak and wet beside her and hopelessly ignorant, having been in the world for only a few minutes. Nothing could be more tempting to a predator, so why would a doe want to do this by herself? Why would she take such a risk unless she had to?

Some animals won't. After hearing this story, my friend Katy Payne, who visited me while this book was in progress, reminded me of elephants she and I had known in the Washington Park Zoo in Seattle. There, alone in a cage, a young primapara elephant labored painfully for most of a day, moaning terribly and very loudly but not delivering the calf. In an adjoining cage, other elephants were also bellowing—her mother and another female. At the time, people knew next to nothing about elephant birth, but at last, almost in desperation, her keepers opened a gate and let the two other elephants join her. Within moments after they hurried in, the young elephant delivered her calf.

I find it hard to believe that any primapara of a social species prefers to labor alone. Looking through the lens of the Kalahari where, in the past, the women gave birth alone in the veld, I remember that the primapara women had their mothers with them when birthing, and were alone only for subsequent births. I recall my own birthing experiences, especially the first, during which I was in no shape to defend myself. I was alone, due to now outdated hospital regulations, right up until the baby was crowning, at which point a man came storming in and slashed my vulva with a scalpel.

I yearned for someone to protect me. As for the deer, it struck me that most of the births recorded in the literature took place on deer farms or in places like deer farms, where the social arrangements of the deer could have been distorted. Dr. Rue's observations—useful and compelling as they are—were of deer in somewhat unnatural situations.

So in the spring of 2008, I kept watch on my field for the Betas and the Deltas. They usually came in the afternoon, the Betas from the woods to the east, the Deltas from the southeast, and they grazed until after sundown. Because a doe who is bulky with pregnancy cannot run at full speed and thus cannot readily escape a predator, fetal fawns gain most of their birth weight all at once, at the very end of gestation. My slender Betas and Deltas seemed not to be pregnant at all until early June, at which point the two mothers ballooned. Their daughters didn't. Day after day I kept watching, expecting to see the mothers slimmed down, but still they kept coming, as huge as ever, eating and looking, eating and looking, going back to the woods in the evening to lie down and chew their food.

It was at this time that the Betas vanished. But the Deltas kept coming, the mother still bulging. One day it rained, but the Deltas came anyway. Toward evening the three of them headed off to the woods, in no particular hurry. Luckily, I was watching them with binoculars, because just as they reached the edge of the woods, I saw the Delta mother pause. Then something huge, shiny, and pink slid out of her vagina. Her two daughters strolled past her and she followed them into the woods.

What had I seen? At first, I thought it was a fawn—the mother had given birth before and might not expect much difficulty—but why, if it was a fawn, didn't the mother then tend to it? Then I thought it might be a placenta, although a doe normally eats the placenta, and if it was a placenta, what had happened to the fawn? I felt I had to know. So without taking my eyes off the place where this happened, I waited until the deer were away in the woods, and then went to take a look. The sky had cleared but the grass in the field was still shining with raindrops, and through the grass each deer had left a highly visible trail by knocking off the raindrops. The Delta mother's trail was in the middle, and I followed it to the exact place where I knew for a fact that the pink thing had emerged. And there, I found nothing. No fawn, and no placenta. So what had I seen? I realized it must have been a gush of sunlit birth water, which the wet earth had absorbed. This meant that not far from where I was standing, the Delta mother was soon to start labor. I left quietly, walking quickly but in a casual manner to show anyone who might be watching that I had business elsewhere, and other things on my mind.

This doe had made no effort to get away from her daughters. On the contrary, she had followed them into the woods. So why is it said that a doe gives birth alone? Some do, obviously, but all of them? What advantage would that give? A companion would pose no danger, and the social groups of other ungulates protect one another, either intentionally, like musk oxen, or by the sheer force of numbers, like wildebeests. Could anything but good come from having a companion or two? Wouldn't a doe in labor be better off with grown daughters nearby, to fight off a predator, should one be drawn to the scene by the possible moaning and the smell of blood? I will always remember the three deer mentioned earlier—who in retrospect were surely a mother and her two daughters—who threatened a coyote with snorts and foot stamping, and I will always remember the demeanor of the coyote, who was half hidden by a boulder but nevertheless was seriously worried by the aggressive deer, and tried to seem unimportant until he could sneak around the boulder and take off. It was thus my impression that the Delta mother, when she went into labor, would welcome her daughters nearby. It certainly seemed, when she followed them into the woods, that she had no objection whatever to their presence and would not have followed them if she did.

By a splendid coincidence, my impression was soon

confirmed with the help of my wonderful friend Sy Montgomery. Sy called one day to say that her neighbor, Hunt Dowse, had witnessed a doe giving birth early that morning in his field. I immediately phoned him for the story, and he told me that he had looked out his window at about five o'clock to see a doe lying in the field with a tiny wet fawn lying near her hindquarters. Evidently she was still in labor, delivering a second fawn. With her, almost frisking at the sight of the first fawn, was her grown daughter, and also a smaller daughter, a yearling. There they were with their mom, keeping her company in her hour of need, welcoming the new arrivals.

What to make of all this information? I'm not sure, of course, but it does seem possible that a deer in her first pregnancy might of necessity give birth alone, while a deer with daughters might do so in their company. The daughters seem to stay with their mothers for as long as they can, because "follow your mother" seems to be deer policy. But it doesn't appear to work the other way, or not strongly—there is no special rule saying "follow your daughter."

Even so, there is more to this question. When I told these stories to other friends, Ilisa Barbash and Castle McLaughlin, I was rewarded with a slightly different story that shows, I think, the nature of the mother-daughter relationship. Castle was doing research on wild mustangs in the Dakotas, and learned that on a nearby ranch when the rancher was away, a heifer began to give birth with great difficulty. (A heifer, of course, is a cow that has never had a calf. Thus, for the heifer in question, the birth was her first.) The rancher's wife did not know how to help the heifer and was waiting for her husband's return. Hour after hour, the heifer cried and moaned, unable to deliver. Then a cow appeared in the distance, coming in her direction. As she came, she had to force her way through three strong barbed wire fences, but at last she reached the heifer's side. Who was she? The rancher's wife didn't know, but late that night when her husband returned, he identified her as the heifer's mother. She had been pastured three miles away.

How did she know of her daughter's difficulty? That's a mystery. Perhaps cattle, like several other ungulates, have an infrasonic component in their calls. Since 1983 when Katy Payne discovered infrasound in elephants, it has been found in the calls of several different kinds of animals, but I'm not sure that anyone has bothered to look for it in cattle. Infrasound has very long waves, and travels far. We humans can't hear it, although the waves travel right through us. The word "sound" refers to what we *can* hear, hence we have named these waves "below sound." That's true for us, and since we consider ourselves to be the only creatures of importance—if people can't hear it, it isn't true sound—we named it accordingly. But it is certainly true sound for those who make it. They can hear it. And even without an infrasonic component, the voices of cattle travel very far, if conditions are favorable. Years ago when our farm was active, we sometimes heard the voice of a bull on a farm at least two miles away through thick woods. Perhaps despite the distance the cow heard her daughter.

What did the cow think she could accomplish with her journey? Obviously she knew about childbirth—she had birthed the heifer. Perhaps she couldn't help with the delivery, but she recognized her daughter's voice, knew she was in trouble, and could protect her daughter from attack, as few things could be more tempting to a predator than a helpless female in childbirth, bleeding and crying.

To me, these stories raise some interesting questions. True, childbirth is exceptionally painful—it is said to be the ultimate pain, the most anyone can feel. Worse things can happen, to be sure, but you can't feel more pain because your nervous system doesn't transmit it. Our nervous systems do not differ significantly from those of other mammals, so this is surely true of other animals as well, especially those who, like ourselves, deliver fairly large babies. But not everyone in childbirth makes loud calls. For some unknown reason I didn't scream or moan when I was in labor, and neither did my female in-laws whose birthings I attended, even though, in one case, the woman in the next delivery room was screaming her head off. We ground our teeth and growled, but we didn't make loud vocalizations. Perhaps this was instinctive. When in pain, most animals—certainly wild animals—are dead silent, no matter how much discomfort they may be experiencing. The last thing they want is to draw attention to themselves when they are seriously compromised.

And what is the purpose of a loud call if not to draw attention? An animal caught by a predator often screams. Why? Because another predator may hear it, and come to see what is going on. People who hunt coyotes, for instance, use a call that sounds like a rabbit screaming. A rabbit would never scream for any reason other than that a predator grabbed it. The coyote thinks that perhaps a meal is in the offing, and visits the scene. Thus the rabbit-call technique merely takes advantage of what happens in nature. If, with luck, the second predator is more formidable than the first, or if he tussles with the first to get the victim, the first may drop the victim, who then has a chance to escape while the two predators are involved with each other. This happens, not all that often, but often enough to make a scream worth trying.

Where certain animals might shriek, the ungulate families make loud, shuddering calls. So perhaps the deer in labor who vocalize loudly are seeking help—not consciously, probably, but in the general, evolutionary scheme of things. This would parallel our kind of screaming. We rarely plan to scream. Instead, we usually find ourselves screaming as does the rabbit caught by a fox. In other circumstances a doe would probably do what all wild animals in pain normally do, which is to keep silent. Yet sometimes a doe in labor calls out. She is in extremis, and Gaia tells her she needs help, so she takes a chance that her mother or perhaps her daughter will hear her before a predator does. Perhaps that's what Dr. Rue's primapara doe was doing. Evidently that's what the heifer did, and in her case, it worked.

A few days later I saw the Delta mother and her two daughters in the field near the pond. The doe was no longer pregnant. Rather, she was standing with her head up and her thigh loose, beneath which something was moving. It was her little fawn, nursing. Her daughters stood nearby. A few hours earlier,

I had discouraged someone from driving a truck to the pond, and was glad I had done so—the fawn was standing just where the truck would have gone, but if the truck had passed there, the fawn would have been hiding. Newborn fawns are so small that they hide under ferns and other low-growing plants, and the driver wouldn't have seen him.

When the fawn finished nursing, he frisked for a moment, and his mother and one of his sisters put their noses down near him. Then they started for the woods with the fawn behind them. They passed through grass up to their shoulders, or long enough to completely hide the fawn, and when they emerged the fawn was no longer following. The grown deer went casually into the woods. The fawn had stopped somewhere along the way to curl up under a fern or in a tuft of grass where he would wait until the others came back for him. They may have done this later, but I didn't see them.

Chapter Six
Fawns

The fawns of the deer family are amazing animals. Just minutes after a fawn is born he is able to nurse, which he does for a bit, crawling up to his mother, then tries to stand on wobbly legs, and soon is able to follow his mother as do the young of the horse and cow families. The mother will lead him away from the birth area for safety, even though she will have licked up all the blood and eaten the placenta and any plants that had birth matter on them—even plants that deer don't normally eat. Still, no matter how clean the birthplace may be, it's better to find a new place, so she leads, and the fawn follows. But then, instead of continuing on as one of the antelopes might do, she stops and somehow persuades the fawn to hide. The fawn creeps under a low-growing plant, curls into a ball that would fit on a salad plate, and stays there. The mother leaves. If she has twins she will lead the other to a different place, as she does not want them together—the separation is insurance against predators. Once in a while a well-meaning person will find a hidden fawn and take him home to feed him cow's milk from a baby bottle, assuming he has been abandoned. Almost certainly, he has not been abandoned, and this is a terrible thing to do, as a person cannot raise a fawn properly, nor is cow's milk good for him. In the unlikely event that he was orphaned, it's possible that another deer will adopt him, as perhaps was the case with the Betas.

But mostly, nobody sees the fawn, no person and no animal. He is the color of the earth, spotted to look like dappled light, hiding in a dappled shadow where he has no outline, and he doesn't move. He also has no odor, as his mother has licked him perfectly clean, nor does he urinate or defecate until his mother returns and stimulates him to do so. She ingests his urine and his feces—a common practice for many animals who have helpless or partly helpless young—and carries them away. Thus, even if the fawn changes place, no odor remains to suggest his existence. The only way a fawn could be more difficult to notice would be if he wasn't there at all.

When a fawn is hiding, no matter how young, he takes responsibility for his own protection and keeps his head up, alert to what is going on around him. If anything comes near, or even if a plane flies over, he curls up tight with his head down and his ears flat, but he keeps his eyes open. If the disturbance gets worse, he enters a state that can almost be called suspended animation. Dr. Rue cites an amazing study by Nadine Jacobsen of Cornell who investigated the physiology of fawns, measuring their heartbeats and respiration as they hid.[1] Jacobsen arranged five phases of disturbance, the first phase being no disturbance. The fawn rested with his head up and his ears forward. His average heart rate was 177 beats per minute, during which time he took twenty-one breaths. In phase two, an observer approached him. He instantly flattened himself, stopped breathing, and his heart rate fell from 177 to 60 beats per minute. In phase three, while the observer sat quietly nearby, the fawn's heart rate crept back up and he began to breathe normally. In phase four, the observer moved about. Again the fawn stopped breathing and his heart rate dropped. In phase five, the observer left the fawn alone. His heart rate rose to 183 beats per minute and he took thirty breaths, almost panting, trying to regain the oxygen the experience had cost him.

This is why hiding fawns are not found often. But they don't hide all the time. They get up and change places, and may do this several times a day. As has been said, a doe often hides a fawn in our field, and one year I noticed a fawn running right and left, bucking, shaking, and running again. In the field there are ground-nesting hornets, and I believe he had disturbed some of them. When his mother came into the field to nurse him, he wasn't where she had left him. She didn't seem at all frantic—it was as if she expected him to move. She poked around here and there and finally noticed him scrunched up in a grassy place, about fifty feet from where he was before. She nursed and cleaned him and both of them seemed calm, as if things were as they should be.

In 1998, however, a fawn in the field was killed by a dog. The dog, a black Lab, belonged to a friend. We happened to be outside standing on the lawn with our dogs. Suddenly the Lab noticed something in the field. He must have known what it was—he took off running with my dogs after him, and his owner couldn't call him back. When he grabbed the fawn and we saw what he was doing, I ran too. He ran ahead of me with the fawn bleating and struggling in his mouth. I ran as fast as I could, yelling at him, hoping he would drop the fawn and I could save it, but he ran into the woods and up a hill and by then was out of sight and so far ahead that chasing him was useless. I went home and he showed up about half an hour later as if nothing had happened. A neighbor then phoned to say that she had seen a large black dog carrying a dead fawn. She asked if it was my dog, and I told her it wasn't. She assumed that he had found the fawn already dead and had simply picked it up, and I didn't enlighten her. What he did with the fawn we'll never know. Perhaps the neighbor caused him to drop it. Strangely, no doe came into the field to search for the fawn, at least not in the daytime. Perhaps she was nearby when the event took place and knew what happened.

Another fawn was killed a few years later, probably by something larger than a dog. Half a mile from my house I found the severed head of a young fawn in the road. I examined it carefully and was amazed to see that the cut was as clean as the cut of a knife, not chewed upon, just sliced, so I assumed that the killer was big with large, sharp side teeth, whether molars or premolars—the teeth with which the carnivores cut their meat. I had no reason to think that the fawn had been hidden near us, and I saw no doe searching for him. If she did, it wasn't in the daytime, so the whole event may have happened far away, or else this doe too knew what had happened.

Sometimes the fate of a fawn is not clear to the doe. One of the saddest things I ever saw was a doe in our field early one morning, searching, searching. Obviously she was looking for a fawn who had been there on her last visit but, it seemed, was there no more. Perhaps a predator had killed it during the night. After a while the doe abandoned all caution and seemed not to notice or not to care that I was watching as she anxiously looked here, looked there, went back and looked again, tried another part of the field, and another, and another, so distressed, so anxious, that my heart ached for her. She finally gave up and went into the woods. But she couldn't believe that her fawn wasn't somewhere, and in an hour or so she reappeared to search all the places she had searched before, as if she might have overlooked something. I saw her again in the afternoon, and again the next day. Perhaps she was telling herself stories of where the fawn might be and how she would find him. In time, her hope faded and she didn't return.

I remembered her in connection with the Deltas. When the fawn followed his family into the long grass but did not emerge, I felt sure he was hiding there. But at that time, strange things began to happen. I saw the Deltas the next day, but not the fawn, and at no time did the mother seem to be looking for him or nursing him. This went on for several days. The mother and her daughters would come into the field, but the fawn never showed himself, and the mother never seemed to be nursing. Something seemed to be wrong.

Then I began to notice that the smaller daughter, the graceful young doe with the pointed face, was staying about a hundred yards away from her mother and sister. If she came nearer, they stared at her. She seemed unwelcome. She felt unwelcome and would stand still and look at them, with her tail pressed, her hips almost tucked, her head up, and her ears forward—the very picture of someone who wants something—but she would not approach them. When the mother and older daughter went into the woods, the younger daughter followed, but cautiously, and far behind them. This went on for several days. Then one day, the smaller daughter came tearing past my office window. I could almost have touched her. She seemed extremely agitated, and stopped to look around the field, her ears far forward and her neck stretched. Then she ran to the place where the fawn had been when I first noticed him. She looked around there for a moment, still very agitated, then ran to the edge of the woods, stopped, and looked in among the trees, then spun around and went bounding eagerly into the woods at the place where the Deltas usually entered, at the deer trail beside the swamp. For many reasons, I felt sure she wasn't looking for the fawn—it wasn't her fawn and she hadn't been pregnant, nor did she have the demeanor of a doe looking for a fawn. She was searching more widely, running here and there rather than stopping and searching, looking far off into the woods rather than looking in the grass. I felt sure she was looking for her mother and sister, but how did she come to be so far from them, and why had they abandoned her so completely? They certainly had seemed to want to separate from her, but why?

I worried about her. I also worried about the fawn, who had not been seen since the day he dropped behind his family in the long grass. Now and then I'd see the Deltas in the field again, but just the mother and the larger daughter. They seemed to have gotten rid of the smaller daughter. Here again, the mother never once seemed to be nursing a fawn, nor did she look for him. I came to the very sad conclusion that he had disappeared, perhaps weeks earlier. The long grass was very near the swamp and the woods, both places that would shelter predators, as was the place where I first saw him.

It seemed as if the mother had willfully cast away her younger daughter as well. Deer live as social animals, yet their rules sometimes demand antisocial results. But what rule might the daughter have broken? Perhaps, like the daughter of the deer in the woods so long ago, she had imagined that her mother's milk was meant for her. Perhaps she kept trying to nurse, and if so, the mother would have had to chase her off. The fawn, if he was living, would need all his mother's milk if he were to survive in winter.

Needless to say, fawns become more vulnerable to predators when they start following their mothers, as the Delta fawn had been doing, however briefly, and which any fawn may do as early as midsummer. In our area fawns are preyed upon by black bears, coyotes, and even bobcats, who have little luck with adult deer. Small fawns are preyed upon by animals such as foxes and eagles and perhaps even fishers, none of whom have the slightest hope of catching an adult deer. But ironically, another of their dangerous predators are dogs. Witness the visiting Labrador retriever. There are always dogs and people don't always contain them, yet if anything captured the Delta fawn, it wasn't a dog, because that year no dogs but ours came anywhere near, and our dogs did not voyage on their own and did not go hunting.

In that case, what was it? Once again, I could only guess. If one's subject animals don't wear radio collars or are otherwise interfered with, guesswork is about the only tool at one's disposal—that and after-the-fact observation, unless one is lucky enough to witness an actual event. So I guessed that if that fawn was lost to a predator, the predator was a bear. I knew for certain that one special bear included our land as part of his territory. He had been injured when he was young, so he was partly disabled, which put him at a disadvantage. For him, a fawn would have been a bonanza.

Here is what happened to the bear. Late one night in the summer of 2002 while this bear was crossing the road in front of our house, he was hit by a quarter-ton pickup truck driven by our friend and neighbor, Don Schrock. Our house is on the east side of the road, and the bear may have been heading for our grapevines.

Only Don saw him. He later said that the bear was about as big as a large dog, and weighed perhaps 150 pounds. Don is an experienced hunter and an expert on the local wild animals, and his description would be accurate. This could mean that the bear was about two years old, or else that he was older but not finding adequate nourishment. It was hard to know. He could have been one of two bear cubs I'd seen at the roadside in the summer of 2000 when on my way home. I stopped the car. One cub was standing on his hind legs with his hand on a telephone pole. His sibling was on all fours beside him. They looked at me with shy, polite expressions. Both seemed to be the wrong size, too big to have been born during the past winter but too small to have been born the winter before, meaning that they may have been about eighteen months old or thereabouts, but were undernourished. Unlike many of the large carnivores, a bear seldom gets to eat a large meal. Instead he must range widely, finding a mushroom here, a few berries there, grubs in a fallen log a bit later, often with a long, calorie-consuming walk between bites, which makes it difficult to gain enough weight to survive a winter of hibernation, let alone to grow. Bears had been extinguished from southern New Hampshire by the early farmers, and had repopulated our area only about ten years before, an overflow from the northern population. Our area is not prime bear habitat.

At the edge of the field behind the cubs was a grassy thicket around a small pool of water, a damp place much favored by animals. Deer hide there, and woodcocks nest nearby. I didn't see the mother bear. Perhaps she was hiding. After watching me politely for a while, the two cubs went into the thicket.

This was where the accident happened—the bear who was hit emerged from that very thicket. The truck hit the bear's right hind quarter hard enough to smash the radiator and front fender with a bang so loud I heard it in my house. The bear bounced off the truck, then picked himself up and climbed a tree. Then he fell out of the tree. Then he scrambled up the bank toward our house and hid in the bushes while Don, who needed an accident report if he wanted to collect insurance, called the police. A police car must have been nearby. From a window I saw the blue lights flashing and ran down to the road, arriving just in time to see the police officer unfasten his holster.

If, as Don was saying, the bear had been able to climb a tree and then climb the bank, it seemed to me he might recover. I told the men they shouldn't shoot him. The men said that the bear was suffering. They also said he was dangerous. He had to be shot, they insisted. I said I wouldn't let them. They told me to go home. I said, "I am home." They told me to go back to the house. I said I couldn't. The officer wondered aloud if I might have been drinking. Don said, "She doesn't drink. She's always like this." The officer had not yet taken out his pistol, but he started to cross the road. So before things could go any further, I scrambled up the bank to the bushes and the bear and told the officer to stay where he was. The men looked at each other. The officer said, "It's not your bear."

I said, "No, but it's my land, it's posted, and you'll need a search warrant to walk on it. Go find a judge," I told him. I also sent a thought-message to the bear. *If you can travel, this would be the time to do so. I don't know how long I can hold them off.* Just then I heard what I'd been hoping for, the bushes rustling behind me as the bear moved away. The two men looked at me for a while, then gave up, got back into their vehicles, and drove off fast to show their annoyance. I waited to make sure they didn't return, and then went back to my house, planning to look for the bear by daylight.

But in the morning Don came by to see if I'd changed my mind. I hadn't, so he made a phone call to the game warden, explaining the accident and asking permission to track and shoot the bear. The game warden gave permission. I said that it was still my land, and I did *not* give permission. Don said he was going anyway. A wounded bear was dangerous, he repeated, and this bear was his responsibility. I said that if he insisted on going, I would go too. He didn't want this, so he left, letting the door shut noisily behind him. Thus it was me and Sheilah, the more cautious of my little cattle dogs, who went to the place where the bear had been hiding to learn as much as we could about his condition. We found no blood and no dead body. That, at least, was good. We then tried tracking. The Kalahari people were the world's best trackers, and I had learned something of the art from them, but the forest floor was a jumble of little plants and leaves with not many places where one could say for sure that an animal had walked. So I relied mostly on Sheilah. She doesn't track as a bloodhound might, nose to the earth, absorbed with what he's finding. Instead, she stays tight against me and bristles when she catches the scent of something scary. She bristled where the bear had hidden in the bushes, and again a few yards to the south, and again farther on as we followed a likely line of travel, and again when we were deep in the woods. She had reacted this way before to bear scent but seldom to any other scent, so I felt fairly sure we were tracking a bear rather than some other animal.

This proved to be true. In the sand by a stream we found two overlapping bear tracks, the left hind foot and left front foot, both set firmly. So he seemed to be getting around adequately, and therefore might recover. Sheilah and I went home. I called the game warden and told him what we'd found. The game warden was noncommittal.

I felt sure we had been tracking the injured bear, if for no other reason than that bears tend to live alone, except for mothers with cubs. That being so, he'd be the only bear. As for saving his life, I saw no reason not to. I don't credit the "wounded animal as bogeyman" theory. During hunting season, the New Hampshire woods abound with wounded animals whom hunters shoot but don't kill, bears included, and nobody is ever harmed by any of them. Nor have I any sympathy for putting an animal "out of its misery"

unless there is absolutely no hope of recovery, and not always even then. In fact, I find the practice appalling. It shows how poorly people of our culture understand other species or even other human cultures. We in the Western world are terrified of pain, but there are people who are not, and many animals clearly are not, and thus are vastly more capable of enduring pain than is our hopelessly soft population. It's not that animals don't feel pain just as we do. It's that they can cope with it and we can't. A person whose hip was smashed by a truck would lie in the road, screaming. The equally sensitive bear climbed a tree to escape from the truck, then climbed the roadside bank to escape from the people. Was that the behavior of an animal who didn't want to live?

But could he continue to move about on broken bones? I thought he could, as I'd seen similar situations. While participating in Katy Payne's research project in Etosha National Park in Namibia, watching at a water source where elephants drank, we had seen a full-grown elephant who had recovered from a broken leg, and thus had been vastly more compromised than the bear, as a bear, like most animals, can walk on three legs but elephants must use all four. Our research team also came to know a very young male elephant who always seemed alone. This is unusual for elephants, particularly young elephants, who normally live with their mothers and other relatives. Perhaps he was the only one in his family who had escaped from a culling operation. Even so, he was a cheerful little guy—for fun, he would chase birds who came to the water, and he even threw some friendly glances in our direction with his trunk held high. One day we saw him walking very slowly and limping terribly, and realized that he had broken his left front leg. Gone was his good cheer—he seemed sad and in pain. An obvious solution to his problem would have been to lie down and wait for death, but this wasn't his choice. He preferred to struggle on, however painfully. We worried about him very much, but he began to get better. A month later, he seemed substantially better, although the two broken ends of the bone were obviously not quite joined and the healing leg was crooked. Still, he was getting around capably enough, eating and drinking, and promised to recover completely if always with a limp. But then one day the park authorities made a rare visit to the park's interior, where they noticed him and shot him dead. He was suffering, they claimed. They had put him out of his misery. Despite his obvious strength and courage they probably thought they had done him a favor. He could have lived for fifty more years if they had not so ignorantly assumed, as many other game managers assume, that they know more about animals than the animals know about themselves. And who will refute their arrogance? The animals can't, of course, and if they could, the game managers wouldn't listen to them.

Here again, I would look through the lens of the Kalahari, and see the hunter-gatherers there, and remember how calmly they dealt with pain, even terrible pain. I remember events in which people were in great pain but seemed to ignore it. If those people had been asked whether they would rather endure the pain or be put out of their misery, they would have laughed at the question. All things considered, this seems to be the way of the natural world. It is certainly the way of wild animals who, if in too much pain, can always invite their own death without any help from us, simply by hiding quietly somewhere until death takes them. Desperately sick animals sometimes do this. But, just like us, they mostly prefer pain to death, so if possible, they struggle on, despite the pain, in hope of healing, and if they heal, as many do, they continue their valuable lives.

That's what seemed to have happened to the bear. I know I saw him in April 2005, walking through the woods on the far side of the pond. By then he was almost six years old, and even though, in keeping with the accident, his right hip was twisted out of line so that he limped like a dog with hip dysplasia, he had managed to grow much bigger. I'd say he had gained about one hundred pounds. The old injury didn't seem to bother him—he moved right along in a fairly normal manner. What better evidence than this to vindicate my demand for a search warrant? I had given him many years of life.

We have a hummingbird feeder and a regular bird feeder, the former hanging from a bracket, the latter on a steel pole, both at the outer side of a long wooden ramp that leads to the kitchen door. The hummingbird feeder had been there for years and seemed safe enough, but every night, I brought the regular feeder into the house so that it wouldn't attract bears. This preserved the feeder until the end of May 2008, when one night I forgot. In the morning I was disheartened to find the steel pole bent double and the bird feeder in pieces beside it. Only a bear could have done this. I happened to discuss the event with Ben Killam, a noted bear authority, and he said, "The bear was hungry." And yes, I'm sure he was. Assuming it was my bear, he was bigger than ever and needed more food than ever to make it through a winter. I didn't begrudge him the birdseed. Instead I bought a new bird feeder and a new pole, and was even more careful to bring the feeder indoors before dark.

That June, the full moon was so bright I thought I'd be able to see animals in the field if any were present. Hoping to catch a glimpse of the Deltas, I went to the kitchen door, a glass door, and cupped my hands beside my eyes to take a look. Surprisingly, I saw nothing at all—just total blackness. This seemed impossible. I looked out a nearby window and saw the whole moonlit scene of the fields and woods, every leaf, every grass blade. I tried again at the glass door, and again saw nothing. As I looked harder and longer, as my eyes got used to the solid black wall, I wondered if an unknown person for an unknown reason had draped a black blanket over our door. But then the blackness began to seem somewhat fuzzy, and I realized I was looking into fur. The bear was pressed against the other side of the glass. We were just a fraction of an inch apart. I turned on the porch light and saw that he was standing on his hind feet with our hummingbird feeder in his hands, tipping it into his mouth to drink the syrup, as a person might drink from a bottle. As to his size, my eyes are fifty-five inches from the ground, and I had been looking straight at his ribs.

He didn't seem to mind the light, but he must have sensed someone near him, because he turned and looked down at me. Our faces were less than a foot apart. Our eyes met. We gazed at each other for a moment, then he dropped to all fours and seemed to float down the wooden ramp to vanish in the shadows. Even with his injury, he was graceful.

I thought I'd just had the greatest experience of my life. This persisted for several days, until I had a greater experience. My husband, Steve, and a friend named Anna Martin were with me when we heard clumping footsteps on the wooden ramp. We thought someone was coming to visit. We went to the door and saw the bear walking toward us. As he came straight at us, we saw that his left shoulder was disabled as well as his right hip—perhaps injured when his body hit the road. He saw us too but didn't mind. Instead he stood up on his hind legs, grabbed the new steel pole of the new bird feeder with his teeth and hands, and bent it to the ground, as a person might bend a paper clip. This dislodged the bird feeder. The bear stepped off the ramp, lay down, and began to lick up the spilled seeds.

We could hardly believe our eyes. It was broad daylight and he wasn't four feet away. He knew we were there, of course—he was looking right at us—but he didn't mind. When his pink tongue had gathered all the seeds it could reach, he bit the bird feeder and crushed it.

Gaia gave us the gift of Fear to keep us out of trouble, but she neglected my husband. He opened the door, walked out, and told the bear to leave. The bear raised his head and stared at him. Anna and I grabbed Steve by the arms, pulled him back in, and slammed the door. The bear lowered his head and went on eating. Then he thought for a minute, as if considering Steve's brief appearance. He stood up and dragged the bird feeder a few feet away. That was better. He could still see us through the window but seemed unworried. He lay down and again began eating.

I'd never seen anything like it. Many a feeding bear would stand on all fours, ready to depart in case of trouble. Was this bear lying down because he felt relaxed, or because of his injuries? Did the damage make stooping difficult? I wanted to photograph the scene so I could consult a bear expert. My camera was in my office, which is in another building. I left the house on the far side and went to get the camera, the bear watching me as I went but keeping his tongue busy. I came back to the house, again in full view of the bear, and took a picture out the window. The bear watched calmly from the corner of his eye. I took a number of photos, but glass was in the way, and after a while I quietly eased open the window just a crack, just so I could take another photo without the glass. At this the bear—who had not seemed to mind Steve's sudden appearance or my travels to and from my office—instantly took fright. He got right up and hurried away, heading for the pond almost by the same route he had taken after the accident, still walking like a dog with hip dysplasia, his left front leg buckling with every step. Without a doubt, it was my bear, but eight years old and enormous.

No seeds were left in the bird feeder, just a few scattered on the ground. He hadn't needed to endure us any longer. As for the amazing sight of a large black bear lying on the ground right by our house, my photos didn't really capture it. I'm no photographer, and the results only show a dark blob against a green background. But a discerning viewer could identify the blob as a reclining bear, so at least the photo confirms my improbable story. And I kept the steel pole. It's a steel pipe, really—the kind used in scaffolding—and in the blink of an eye the bear bent it double.

When I was a child, I learned of a family that built a low platform near their house, and on the platform put various kinds of foods, from kibbles and meat to vegetables and birdseed. These foods attracted wildlife, which the family would then watch through a window. To me, this seemed ideal. I dreamed of doing likewise.

But I never did. Instead, I mostly heeded the warnings of New Hampshire Fish and Game, denying myself for several reasons, not the least of which is that if a bear believes that people will feed him, he visits people's homes, which can result in his death. For instance, a man in a neighboring town saw a bear on his porch. He thought the bear would harm his wife, or so he told the game warden after he had shot the bear. The game warden told him that the bear was after the bird feeder on the porch, not after the wife. As the game warden put it, the bear thought the bird feeder was a bear feeder. Tragedies of this kind occur time after time, and I don't want to cause any of them. After I realized that taking a bird feeder indoors at night was insufficient, I resolved not to maintain a bird feeder until the bears were hibernating. Even in the autumn, a bear doesn't get enough food from any one bird feeder to make much difference to his well-being, and other foods are available for the seed-eating birds.

Yet it was the behavior of my dogs that spoke most strongly of the folly of encouraging bears. When the bear departed after emptying our bird feeder, I went outside to salvage what was left of it, and my two dogs came with me. Miraculously, they hadn't learned of the bear when he was present. But the moment they stepped outside they caught his scent. Up went their lips, fur, ears, and tails, and they rushed off after him, barking so loudly they couldn't hear me screaming for them to come back. They had devoted their adult lives to keeping intruders off the property. Now a monster had trespassed, and they felt they must confront him. Gone were the wonderfully cooperative dogs who always came when called. They were all the way to the woods before I realized that calling them was futile. I began yelling NO, and that alone stopped them. Still bristling, they briefly considered what I'd said—the angels—then reluctantly came back to safety.

What if they'd caught up with the bear? Considering the degree of their excitement, they probably would have attacked him. Another bear might climb a tree, but this bear was disabled. If he lacked a choice, he might have had to fight them. The outcome of that would be quite clear—he was ten times bigger than they were. Many dogs have been killed by bears, including the brave dogs that are used to hunt bears. As for my dogs, they go everywhere I go, so I saw myself leaving my office at night with the dogs beside me and a bear somewhere nearby in the dark. The dogs would catch his scent and rush at him before I could stop them, and that would be the end.

After that, I made sure that nothing was left to attract the bear, including the bird feeder. Even so, early one morning in mid-July we looked out the glass door in the kitchen and saw the same bear on the lawn, coming toward the wooden ramp, just as he had come the first time. I rushed around shutting the other doors to keep the dogs in, and the commotion scared him, so he departed, taking the same direction he had taken before, off toward the pond, and on the way he again passed the place where the Delta fawn had been hiding. By then, it seemed, he had passed this way so many times that he could have worn a path. Did this explain why the Delta fawn was missing?

I went there later to look for blood or hair, but I didn't find much. I did find deer droppings and a few deer hairs, but these looked like the winter hairs of an adult, which would have fallen out naturally. A decent sense of smell would have helped, but mine is minimal and I had no way to interview my dogs. I didn't like to think that the fawn had been killed, but there it was—no Deltas, no fawn, and a disabled bear in the area. To find the silver lining of what seemed like a very dark cloud, I thought that the bear would have had a good meal. The food supply of our area did not increase in size, but he did; each year he needed more food than he had needed the year before. To find a fawn would help him.

On the morning of August 3, to hugely complicate the question, a partly grown fawn whose spots were fading hurried past my office window on the short grass between my office and the field—the first deer of any kind I'd seen for weeks. He had come from the south, the direction of the bear trail, also the place where I'd seen the Delta doe with her newborn fawn, and he was heading for the north side of the field where in the past, the Deltas slept in the grass. He was alone.

Who was he? Could he be the Delta fawn, now more than two months old? Could his mother have fed him only at night when I couldn't see her, as a doe will sometimes do? Wouldn't this mean that the seemingly endangered fawn as well as the disabled bear were both surviving?

I wanted to know but didn't expect to. This was a question from the Old Way, the life of the wild that we see in small pieces. I could do no more than the bears and the whitetails do—keep looking and listening for more information. But in the summer of 2008, this had been difficult. In May and June, the weather pattern that in winter brought snow almost daily then brought rain almost daily, and the flourishing grass had grown so long that it was up to the shoulders of a deer. The best I could do was to search the field for patches of flattened grass that deer make when they lie down. I found many such patches on the north side of the field—the place where, if deer were there at night, we would see their eyes shining in our headlights—as far from the bear's pathway as they could get without going into the woods. Some of the patches were big and the bent grass was somewhat loose, as if an adult deer had been lying down briefly. Other patches were small and the grass was packed down very tightly as if a fawn had spent hours in them. Thus, to my great surprise, the Deltas seemed to have been there all along if only at night, perhaps having moved their fawn across the field when they noticed the bear's presence. Perhaps this was why the youngest daughter of the Deltas was searching so earnestly. Obviously, her mother and sister were not where she expected them to be. Perhaps the doe moved the fawn in her absence. Since the flattened-down grass showed that they were using the field, I concluded that she had found them.

One question remained. Why did the fawn walk past my office? The third of August was a very hot day. Since the Delta doe didn't nurse him in the daytime, perhaps the fawn became thirsty. Perhaps he had gone to the pond for a drink of water and was on his way back when I saw him. But then again, perhaps not. That's one of the troubles with wildlife observation. You find questions you cannot answer, and mysteries you cannot solve.

Chapter Seven

Drivers, Hunters, and Their Prey

Deer are most vulnerable as fawns, of course, but they face danger throughout their adult lives too. Most of the danger comes from us. And to a degree, vice versa. An important cause of whitetail death is vehicle accidents. These don't compare to deaths by hunting, but the number is large just the same. Many people are also killed in these accidents. In fact, deer are rated as causing more human deaths than any other animal, with bees causing the next most. That doesn't really make sense, as bees mean to sting you and a mountain lion means to kill you, but a deer doesn't mean to get hit by your car. Still, it's a statistic.

These accidents are unfailingly horrible, even if no human being is hurt. Too often the deer is not killed outright, but just seriously injured. Deer hit by cars are not given a chance to recover on their own as was my bear. They are invariably shot by the police or by some armed bystander, as were two deer whose accidents it was my misfortune to witness.

The first of these took place on the road I take to go to town, where a doe was hit by a car. This was the doe whom I thought might have been the mother of that mysterious third deer of the Betas, the little fawn who might have been adopted. When I arrived at the scene, she was awaiting her fate at the roadside, not struggling, looking at the small crowd of people who had gathered to observe her and were busily telling one another not to let the children who were present witness the sad scene. The children were drooling with fascination and had no intention of not watching.

The doe's legs were broken. A doe cannot walk on broken legs, obviously, and I believed that her destruction was inevitable. But did we need to terrify her, as the onlookers were doing? I persuaded a few of them to give her some space. Then I learned that our neighbor Don had seen the doe on his way into town and had turned back to get his rifle. Soon he returned, and to my horror, he and another man grabbed the doe by her head and her broken front legs and started to drag her into the bushes. Why? Because they wanted to spare the curious children the sight of Don shooting her. To me, the sensibilities of the children were vastly less important than the terror and suffering of the doe—just because animals can endure pain does not mean we should inflict it on them—and I did my very best to persuade the parents to remove their children and begged the two men not to drag and thus torture the doe but to shoot her where she was. I was unsuccessful. The two men dragged the doe about fifty feet into the woods, bumping her over every obstacle, making no effort to ease her pain or to move her carefully. Then they shot her. So the children saw an animal being tortured if not shot.

The second such event I witnessed was just as bad, perhaps worse. In the middle of the day, an antlered male whitetail was hit by a car in the eastbound lane of Route 101, New Hampshire's east–west highway. This seemed especially tragic, as the deer had come from a fairly open area. If the driver had been paying better attention, he might have seen the deer at the roadside. By the time I came upon the scene, people from other cars had gathered and ambulances and police cars were present, lights flashing. Even the damaged vehicle was receiving assistance—a tow truck was hooking on to it. Fortunately, the driver and his passengers were not hurt.

The deer was, though. He was right in the middle of all this, very much alive, terrified, and struggling to raise himself on his broken bones. He kept collapsing when he tried to stand, but he could hold up his head—he was desperately looking at the woods where he had come from, woods that would hide him if only he could go back. No one was paying any attention to him, knowing that the police would shoot him when they got around to it, probably when all the people were safely out of the way.

I have never seen anything so all alone. I could hardly imagine what it would be like, to be so utterly alone, suffering, unable to stand, surrounded by a crowd of some other species who were busily providing themselves with every kindness, every form of care, even caring for a vehicle, and stepping right over me to do so because to them I was nothing at all. I'd be alive, but my life would be over. The deer saw this too. I left before they shot him.

I'd rather die from a bullet than a car accident. I'd also rather be shot than experience the terror and pain of being herded into a slaughterhouse and clumsily hit with a stun gun or butchered alive if the stun gun wasn't working well, as is often the fate of the animals we eat. Thus I have no real complaints against deer hunting. I'd rather be a deer living free in the wild, and suddenly drop dead from a bullet in the heart or brain, than be a cow or pig or sheep in the agricultural industry with slaughter in my future. Wouldn't we all?

So I won't repeat the reasons why hunting is good—population control, no other significant predators than us, etc.—because everybody knows those reasons. But I do believe that tree huggers like myself overstate the reasons why hunting is bad. Hunting is cruel and dangerous, we say. Also we can't take walks in hunting season, not even if we wear orange vests, because bullets are flying and we'll be shot. Or so we think. One year I did a little research to find out how many people were actually shot by hunters, and learned that no people were shot, not even one, not that year and not the year before, although hunters had been out there by the thousands. Well, one man was shot in the leg with buckshot, but he wasn't hurt badly, and certainly not killed. And this, it seems, is normal for New Hampshire. So much for the complaints of we tree huggers.

Why the impressive safety record? The State of New Hampshire requires every hunter to be licensed, and to get a hunting license, one must either produce a valid license from another state or pass the course on hunting sponsored by Fish and Game. In 1998, I signed up for the course. My friends and family were dumbfounded, but my reason was fairly benign—I often wrote about the dog and cat families, predators all, and eventually realized that since all of these were hunters, their evolutionary focus was something I knew little about. Better to learn more before writing more.

The course was fascinating. Most of it was devoted to the safe use of firearms, but one also had to learn the rules. You can't hunt at night, for example, except you can hunt coyotes at any time during their mating season, from January through March. All you need is a license and a gun. This was the one question I missed on the final exam. I was already so horrified to have learned that coyotes were exempt from any meaningful protection that I assumed you didn't need a license. Even so, in a class of about thirty people, most of them macho types who were surprised to find a tree-hugging grandmother in their midst, my grade was 98 percent, the highest in the class by far. Oh, the good feeling. I was issued a hunting license and have it to this day. When you pass the age of sixty-eight, as I did eventually, you are given a special permanent hunting license.

One thing the course did not teach was how to hunt. Whatever information I acquired on the subject I got from watching animals hunt and also from our neighbor Don, who had Micmac ancestors and was a profound observer of the natural world. At the time, he lived in a cabin on the mountainside, from which he kept an eye on the local wildlife and was kind enough to alert me if an animal of interest was heading my way. I would watch for it. We also searched the woods individually and together, and remembered what we saw so well that once in 2004 we found ourselves discussing some unusual droppings we had noted in 1997. We both knew exactly which droppings. They had been large and oval shaped, and Don, with hunting forever on his mind, more interested in deer sign than any other sign, took them to be those of a large buck. I thought they were those of a porcupine. In the intervening years Don had been mentally reviewing the droppings and had revised his opinion of the type of animal that left them. By 2004, his wishful thinking had evaporated, and he agreed it was a porcupine. A big porcupine, but a porcupine nevertheless.

Don was such a good hunter that for him, the hunting season often lasted less than an hour, or it did if he stayed in New Hampshire. He also had a Maine hunting license, and sometimes he would hunt on the Penobscot Nation as the guest of his Indian friends. In New Hampshire, the season opens half an hour before sunrise on a day determined by Fish and Game. A few minutes later the people in our neighborhood would often hear a single gunshot and we would look at each other, knowing that Don got his deer. We would also know he did it with a muzzle loader, not only from the sound of the gun but also because a week or two must pass before Fish and Game allows hunting with a regular rifle, and Don likes to get his deer early. In all the years we've known Don, I remember only one when he didn't get a deer. He is a carpenter and contractor and also a licensed guide, and that year was very busy. Otherwise he got one or more deer annually, and of course he used the meat. In fact, when he wasn't working or hunting, he was fishing, also very successfully, and he fed himself and others, us included, with his bounty.

An interesting fact about hunting is that unlike other sports, nobody knows who the stars are. The only acknowledged measure of success seems to be the size of the trophy, which is usually determined by the standards of the Boone and Crockett Club, named, of course, for Daniel Boone and Davy Crockett, founded by Teddy Roosevelt, and dedicated to big-game hunting and the assessment of trophies. Thus whoever gets the biggest buck is almost by definition the best hunter.

This of course is far from true. In reality, America's best hunters are people like Don, out there without store-bought aids and gadgets, knowing where the deer are and what they're doing, moving silently through the woods in the manner of the big cats. Any other style of hunting, such as baiting one's game or using dogs or complicated gadgets, is something like fishing with dynamite and thus is not a measure of skill.

I realized the scope of Don's abilities one autumn night when he was giving me a ride home from town in his truck. No other cars were on the road, but to my surprise he stopped suddenly for no apparent reason at the place where, years later, I saw the Tau deer crossing. Don said that a whitetail buck was nearby. He had caught its odor. And he was right. Even with my limited sense of smell I eventually caught it—the musky odor of an ungulate not unlike the marvelous scent of cattle in the barns of my childhood. Why a large buck and not some other deer? Because the rut was starting, and a mature buck has a characteristic scent. Don had caught it through the window of his truck while moving along at forty miles an hour. The only other people I knew who might have done this were in the Kalahari.

America's very best hunters are not people. They are animals that range in size from shrews to mountain lions. But Bushman-grade hunters such as Don approach their abilities and use the same skills, or many of them. The first time I went hunting with Don, he noticed the scent of a deer wafting from some bushes we were passing. He quietly pointed this out, and having been told what to expect, I too caught the scent. He eased off through thick brush in that direction, quiet as a cat.

Gaia put the will to hunt deep into our psyches—there's nothing like it. I followed Don step by careful step, eyes wide, ears open, hardly breathing, but so far behind Don that I could see no more than a part of his jacket through the leaves. Then suddenly I knew he saw something. I don't know how I knew this—I just did. I froze. I saw movement as he raised the rifle, then saw him turn, the rifle down. Through the leaves he had seen the hindquarter and antlers of a six-point whitetail buck but not clearly enough for the shot. The buck moved off, and we moved after him until we realized he must have circled around and was following us, feeling safer, of course, if he knew what we were doing. Since he probably knew all about us, we seemed to have no chance of bagging him. We went home.

I recall that hunt as an experience of the utmost intensity. Perhaps extreme intensity can cause ESP, which was how I seemed to know for sure that Don had seen something. There certainly was no standard explanation. Even so, I hesitate to offer one that I hardly believe myself. But then, the life of the Old Way overflows with phenomena that we don't quite understand, and we must accept them as such when we encounter them.

Don and I went hunting again later, in the woods on the west side of the road. There in an open grove of trees we saw a spiker—a fawn of the previous spring with small, spare antlers. He ran. Don shot him through the chest, but the deer kept running. We tracked him for half an hour, then I had to go—a board on which I served was having a meeting I had to attend—so

Don, an excellent tracker who would never abandon a wounded deer, went on alone and found the young deer lying dead on the far side of a stream.

In general, each hunter is allowed to kill one deer, and each hunting license includes one tag that the hunter must fasten to the deer as soon as it's dead. Don should have tagged the spiker, but it was an inconsequential deer and he wanted to get a bigger one, and as I was leaving he wondered aloud if I would tag it, so I gave him my tag. (He used his own tag later on a huge six-pointer whose head he mounted on his wall.) I wasn't sure if you could legally give away your tag, but the hunt we'd just accomplished was more for me than for Don, and I certainly owed him for the experience. I didn't see that it mattered all that much whose tag was on a deer, and hadn't fully realized that I would become the named hunter of the spiker.

By the time I returned from the meeting, Don had dragged the deer out of the woods. Within twelve hours of killing a deer, one must register it with Fish and Game, so Don put the carcass in his truck and we took it to the nearest registration station—a sporting goods store in a neighboring town. As usual in hunting season, a group of men had gathered there to view the deer brought in by hunters, and when they saw me at the desk showing my hunting license and supplying the particulars, they eyed me with suspicion, then eased outside to look at the deer. These men of course were hunters themselves. They didn't think I'd shot it, especially not with a muzzle loader, and most especially since a guy like Don was standing beside it.

So they drew me into conversation, hoping that I'd give myself away. One of them asked where I'd shot the deer. A common answer to this question is "In the woods," as many hunters don't like to reveal the good places to strangers. I said, "In Peterborough." They tried again. "It was standing when you took the shot, was it?" one asked. It wasn't standing, of course, as these men well knew from looking at the carcass. If the deer had been standing, the bullet hole through the skin would line up with the hole in the muscle below. But it didn't, because the deer's reaching legs had stretched the skin of his chest forward, and after death the bullet hole slid back over intact muscle. It is of course much harder to hit something in motion than something standing still, especially with a heavy, awkward muzzle loader, but I had no choice. I said the deer was running. They still didn't believe me, but they had failed to trick me and they seemed a bit grumpy. I felt I'd won, and added a flourish to make them more grumpy: "Lead them a little when they're running," I advised.

I may have outfoxed them, but not the game warden, who was on the phone when I got home. He didn't think I'd shot the deer either, and he knew I hadn't after he tempted me into talking about the gun and I told him I didn't like the powerful kick. I'd never fired a muzzle loader, and had no idea that they don't kick. My fraudulence was revealed. But this wasn't exactly evidence, so there were no consequences to my crime. The game warden, a very fine man named Craig Morocco, is much admired by sportsmen and environmentalists alike, and I regretted my attempt at deception.

I also regretted the death of the deer, whose skin now hangs on my office wall and whose meager antlers stand above the dog door in my office, an echo of a larger pair of antlers that were given to me by someone, and that stand above the regular door. As a fawn, the young deer had survived his first winter but hadn't had time to learn about hunters. Before he saw us, he had never seen a hunter. He had not been sufficiently cautious, probably because for the first time in his life he was experiencing the rut with all its distractions and desires, and when he saw us he ran in the direction he was heading. That brought him right in front of Don. A more experienced deer might have wheeled and run the other way, which would have brought him into heavy cover.

Did I learn how to hunt? No indeed. I had to hit a target with a .22 to pass the firearms portion of the hunting course, and if I say so myself, I'm not a bad shot, because when I was a kid my dad taught me and my brother how to shoot a .22. But I have no wish to fire a heavier weapon, or indeed, any weapon. As I seem to have told the game warden, I don't like the kick. More importantly, I have no wish to take a life, although a hunt itself is so all-consuming that one forgets about things like that. I didn't learn how to hunt, but I did learn how it feels to hunt, which was all I really wanted. And I believe I can say with confidence that hunting with a camera or noninvasive wildlife viewing is nothing at all like real hunting.

As a girl, I used to stalk deer in the fields by crawling through the grass on my belly, and often got quite close before they saw me. If I had been a Bushman with an arrow, I could have taken a shot. I never had such thoughts at the time—I just wanted to be near a wild animal. I wanted to be part of the natural world, to belong to the Old Way even before I knew about the Old Way. Out in the woods on that first hunt with Don, I found I had opened my mouth so I could hear better and taste whoever's scent was in the air. My eyes felt hard. My pulse was racing. I was almost not breathing. I was in the moment—strongly, strongly in the moment. That's what hunting feels like. I looked like and felt like what hunting animals look and feel like. Others who live in the Old Way know all about that.

How do the good hunters learn to hunt? Looking through the lens of the Kalahari where information was passed from generation to generation by shared experience, especially information about hunting, I wondered if some of us in the so-called industrialized world learned to hunt in a similar manner. So I did a little research, interviewing about fifteen hunters from different parts of New England. Most of them didn't know one another and had no contact with one another, and I met them more or less at random, so they could not be considered a group. But all had one outstanding similarity—all had learned to hunt from an older person, often a father or an uncle who in turn had learned from a father or an uncle. Each hunt is unique, with its particulars depending upon the victim, not the hunter, and this would seem difficult to learn from videos or books. I was told of one man, the father of a friend, who hunted every year but had read somewhere that he should smell terrible in order to cover his natural odor. So every year he went hunting unbathed and wearing filthy clothes, and his hunts were always failures. He never once managed to shoot a deer. Better that he had learned from an experienced hunter who knew that deer are extremely sensitive to odor, and that you cannot mask your human odor by making it stronger and worse.

But there was more, again thanks to Don. After our second hunt in the woods, when I came back from the board meeting and saw the carcass of the spiker, I noticed that the testicles had been removed. I asked why. Don said he always did that—he put the testicles in a tree. Why? Because they flavor the meat. But why the tree? Because the scent would spread out to attract other bucks where hunters could find them. Don had yet to use his tag, so perhaps in this case the plan had merit, but always? Does one want other hunters hunting the deer that next year one might hunt oneself? Don certainly didn't, and didn't quite have an answer. He hung the testicles in the tree because the man who taught him how to hunt had done so.

This too seemed like an echo of the Kalahari. The reasons that one did certain things were lost in time. Such was the Old Way.

Interestingly, many of the other New England hunters whom I interviewed also put the testicles in a tree. They too were not sure why. They seemed not to have wondered about it, and were kind of amused when asked. Hmm. Gosh. Why is she asking about this? And then, like Don, they produced a few unlikely and quite different explanations. But all agreed that the men who had taught them how to hunt had done so, and one suspects that the elders of these men had also done so. To me, the custom had a Paleolithic air. Who knew how long it had been going on? Had it anything to do with masculinity? Women hunters are a recent phenomenon, whereas men have been hunting big game for at least 35,000 years. Are some of today's hunters the most recent link in a chain that reaches back to our ancestors, perhaps even to our savannah ancestors? Have older men been teaching younger men on through time, right on up to the skeptical men in our local registration station who correctly assumed that I couldn't hit a deer if it was running? Those men knew more about hunting than I may ever know, and probably from an early age at that, which surely is why it bothered them to see me claiming a carcass. Even if they read these words, which I'll bet they don't, they wouldn't upgrade their opinion of me, because the hunting knowledge they accumulated did not come easily, and my anthropological approach doesn't count.

The Old Way of passing information is, of course, used by all social animals, including the deer, and is why a population of animals in one place may behave very differently from those in another place. My husband and I once hiked in an enormous state park in the Blue Ridge Mountains and were amazed to come upon a group of deer on the trail, a few of whom simply moved aside to let us pass, as people might do in the aisle of a supermarket. They had no fear. We could have touched them. They also had no predators. The wolves and the mountain lions had been eliminated long ago, and human hunting had been banned for many years. The deer population had lost all knowledge of predators. For information to pass in the Old Way, a young, naïve deer must be with an experienced deer when the informational event occurs. Years passed without events, and the skills of detecting a hunter, hiding from a hunter, running from a hunter, stalking a hunter from behind to keep him in view, were all forgotten. The deer on Blue Ridge Mountain could not have been less concerned if we were trees.

Not so the deer in New Hampshire, who have been hunted by our species since the Paleo-Indians arrived. In those days, whitetails escaped from hunters by running away, just as they had been running from their other predators, all of whom were present while these deer evolved, which is why deer can run at forty miles an hour. Running would not have helped much against wolves, who themselves are long-distance runners with phenomenal endurance—deer sometimes escape wolves by going into ponds where the deer, being taller, can touch bottom and the wolves, unable to touch bottom, must swim—but running would have helped against hunters with spears, and also would have helped somewhat against bow hunters as soon as the deer learned the flight-distance of an arrow.

But running doesn't help as much against a hunter with a rifle. The advent of modern hunters and hunting laws created new situations for the deer to cope with. They may have thought they knew about our species from dealing with those who called themselves The People of the Deer, but the newcomers from Europe were not like the Indians. If at first they were not quite as dangerous because they lacked the skills, they became more dangerous later, with modern rifles. And by then, the Indians had rifles too.

Game laws have existed in the British Isles since medieval times, laws that became increasingly complex and carried serious penalties, but when the early settlers came to the New World they brought the game laws with them only as memories, probably unpleasant memories, since European game belonged to rich land owners, and the early European immigrants were more likely to have been the poachers than the landed gentry. In 1646, Rhode Island tried to resurrect some of the game laws, and other states and colonies eventually followed suit, but no one paid much attention to them. By the middle of the nineteenth century, remorseless, nationwide, year-round hunting had all but eliminated every kind of deer, from the moose in the North to the mule deer and the elk in the West to the whitetails in the East. Only the caribou remained, too far north to be available to most hunters. Dr. Rue points out that during the late 1800s in Massachusetts, New Hampshire, and Vermont, deer were so rare that the sight of deer tracks in the snow caused headlines in the local papers. We had killed off the deer before they could figure out how to elude us.

It's said that there were once about thirty different subspecies of whitetails, each adapted to a different area or climate. The most northern were *Odocoileus virginianus borealis,* or northern Virginia deer—those best adapted to places like New Hampshire and southern Canada. Farther north, whitetails don't survive. In much of Canada, moose represent the deer family, and north of them, caribou. However, when deer in the United States were facing near extinction, game biologists and others began to subject them to management. It's probably safe to say that no other mammal in the world has been so heavily managed. And game management, being geared to the sport of hunting, wasn't always the science that it is today. Just as fish bred in hatcheries are regularly transported to various lakes where they are dumped into the water to be caught by sportsmen, so, in the not-so-distant past, deer were taken from the few areas where they survived and transferred to other areas, either to increase a population or to "improve the breed," all for the benefit of hunters.

With so much interference, the subspecies mingled or the indigenous deer were replaced by deer from completely different environments and thus could be poorly suited to the new place. Imagine, for example, deer from Florida being transplanted to the harsh New England climate, where the rut is timed so that fawns will be born in the spring. Florida deer have fawns throughout the year, so their rut would not be set for November. Eventually, nature would sort this out, but until then fawns could be born during snowstorms with no chance to survive. Imagine too the relatively robust northern deer being taken to the South. They would be bigger than the locals, with heavier hair, thus far less able to withstand heat, but their males might triumph over the local males for females, and within a few generations the characteristics of the locals would have faded forever with all the advantages that enabled their subspecies to thrive in a hot climate. The present New England whitetails were introduced in the middle of the nineteenth century, who knows from where? An interesting feature resulting from transplantation was that sooner or later, most new populations adjusted more or less, but the characteristics of their subspecies were erased. According to some authorities, only one or two subspecies still exist as such. All other whitetails are said to be "generic."

When meaningful game laws were finally enacted and enforced, hunting was limited to a few weeks in the fall, daytime only. But as recently as 1935, when my father started his farm in New Hampshire, deer were still rare. The New Hampshire farmers never liked wild animals, including deer, except as venison, because wild animals ate their livestock or their crops. Long before the 1930s the large predators had been destroyed, the fur bearers had been trapped, and the hawks and owls had been shot out of the sky and nailed to barn doors as a warning to others. Not until the 1960s, after most of the farms had gone out of business, did the New Hampshire fauna even begin to return to its original condition and by then the deer knew about the game laws. They knew when the hunting season began and when it ended, and then became less fearful of people especially after dark. In other words, they have learned our present customs and have adapted to them, as they had adapted to the Paleo-Indians before us, and to the wolves and the mountain lions before that.

Male deer are hunted more often than females, and therefore are more cautious. In our area, they normally stay in deep woods in the daytime, and as has been said, they seldom flag their tails because the white is so conspicuous. Not only do they want to avoid a hunter who may be stalking them, but they also want to conceal their very existence. This is their wisdom.

Dr. Rue provides statistics: "Where there is no hunting, the ratio is 100 bucks to 160 does. Where there is hunting of either sex, the ratios do not change greatly: most hunters would rather take a buck than a doe. Where only bucks are hunted, the adult bucks comprise 10 to 15 percent of the herd, adult does about 50 percent, and the fawns 35 to 40 percent."[1] Better to be born a doe.

In our fields, I occasionally see male deer during the rut, when they come into the fields if females are present, and spar with one another, as if they had forgotten all about our species, at least for the moment. As has been said, in keeping with the general pattern of the deer family, young males don't stay with their mothers and sisters. Instead, they gather around older males, and thus can learn from them. They learn, but they are not protected—when on the move the older males let the younger males go first, so if somebody gets shot, it's one of the youngsters and not them. The older a deer becomes, the more he knows about avoiding hunters, and because each year a deer grows ever larger antlers, some of the survivors have eight or ten points on their antlers, and some have even more.

But such deer are rarely seen. They move about cautiously at night, and by day they hide in the deepest woods, in swamps and other dense cover. Some people call them "gray ghosts." During the time that I was talking with any hunter kind enough to answer my questions, I sometimes heard references to these gray ghosts—someone would mention a rumor that a really large buck might be in such-and-such a place, and all the men present would listen, alert but silent, perhaps planning to go there for a look. I'd realize that somewhere in the deep woods a solitary buck who so far had managed to elude human hunters was hiding, but hunters were talking about him, hoping to find him and kill him for the sake of the antlers that he could not help but grow. The older he got, the less were the odds of his escaping. In a way, he was doomed, and if he didn't know all this precisely, he had a sense of it.

All this, I think, is very important. We think of deer as escaping by running, and of course they do. But they also hide, and this to me is very interesting. Trying to hide would be the worst possible defense against a wolf or a mountain lion, if only because deer are themselves olfactory animals and exude scent to inform other deer. Cats are reasonably good at detecting odor, and the dog family is exceptionally good at it. Hiding would also be a poor defense against the precolonial Indians who, in the manner of the Kalahari hunters, knew more about the woods and were better hunters than most of us could ever be. Yet hiding is an excellent defense against modern hunters, out there with semiautomatic rifles and other specialized equipment.

Probably there has not been enough time for evolutionary pressure to create an innate adjustment to the rifle. Thus the change is almost certainly a cultural development that the deer have figured out for themselves. Young deer learn from older deer how to escape from human hunters, and pass the information later to next year's fawns. Thus the practice of hiding may have developed specifically as a defense against hunters with rifles. Did deer across the country figure it out individually or spontaneously? Or did those who survived by hiding have time to teach enough of the others so that the practice could spread? Nobody was keeping track, it would seem, hence as far as we are concerned it's one more mystery of the Old Way that we may never solve.

Deer can hide extremely well. A faint, quick movement such as the flick of an ear in the bushes is the kind of clue that good hunters look for. Deer know this and are able to stay absolutely motionless for hours. If the deer who enter the woods from my field were any example—the deer who knew at exactly what point they become invisible to an observer—

they are excellent judges of cover, and may hide standing up or lying down, depending upon what is shielding them from a potential viewer. It must take nerve not to panic and run while a hunter creeps toward you. But even then they stay perfectly still. Dr. Rue shows a photo of a whitetail in a snowy field hiding behind a tuft of dry grass and saplings.[2] Lying down with his head low, his tail folded to conceal the white, his impressive eight-point antlers almost invisible among the tuft of plants that hides him, he looks like dry grass, and even in the photo with a caption that says what one is supposed to be seeing, one can scarcely find him. His antlers blend with the branches above him, as both have the same thickness and kind of curve, and his coat color blends with the dry grass. As long as he didn't move, he would be invisible. A deer such as he must know exactly what he is doing, including what he looks like, to be able to choose such effective camouflage. Perhaps the deer who hide remember their time as speckled fawns, curled up in the shadows of ferns.

Dr. Rue tells of experimental hunts[3] conducted by several different states in fenced areas perhaps a mile square in which a number of bucks are imprisoned. Biologists in observation towers watch what these bucks do while hunters try to kill them. To learn how these deer manage to escape—information of great value to hunters—is essentially the purpose of the experiments. But if hunting imprisoned deer might seem easy, it is not. Dr. Rue tells of an experiment conducted at the Cusino Wildlife Experiment Station in Michigan. Six experienced hunters hunted for four days before any of them even saw a buck. Not until the fifth day did a hunter manage to kill one.

These deer knew they were being hunted, especially when they saw the hunters sneaking around, and they knew they could not leave the area because of the fence. According to Dr. Rue, some "crawled into thickets or under blowdowns, or got behind or under cover that looked too skimpy to conceal a deer."[4] There, they stayed perfectly still while the hunters walked past them. Some stayed in hiding for a day at a time, just as they might have done when they were fawns. Others would move after a hunter had gone by, perhaps to find a place that seemed safer. In the Michigan experiment, the ground was covered with snow, so the deer who moved left tracks that the hunters could follow. And yet they escaped, because they knew about their tracks and would circle around to watch their own trails to see if anyone was following them.

The account doesn't say if any deer survive these hunts. I'll bet they didn't. After all, the idea was not to see if any deer survived, but to see how long it would take the hunters to kill all of them. Wildlife experiments being what they are, the deer were probably doomed from the start.

But they wanted to live. I know that from deer-feeding.

Chapter Eight

Our Place in the World

Everything wants to live. Everything wants offspring that live. This book is about deer, but watching them inevitably, sometimes serendipitously, led me to a few of the many lessons available from the millions of life-forms who share the ownership of my land. They own it by rules established by their species, just as my ownership is established by a document in the Hillsborough County Registry of Deeds. The bobcat on the slopes of North Pack stakes his boundaries with scats in important locations. A black walnut tree planted in the woods by my father oozed out a toxin to keep other plants from growing nearby and made for itself a little clearing of its own. The tree didn't live, but I successfully use its strategy—my land is posted. Of all the lessons I have learned from the natural world, the most compelling is this: thousands of different kinds of us are here, doing what we must to meet our basic needs. Our methods are different, but our object is the same.

So I try to keep my eyes open for my co-owners. Often I fail to note them because I'm too focused, either because I'm looking for deer or am thinking about them. The great naturalist Olaus Murie had something to say about this mind-set: "Perhaps it is better not to specialize too much," he wrote. "The bird enthusiast can add to his enjoyment and understanding by some interest and skill in reading the record of mammals. The naturalist goes forth to enjoy what he can find, be it bird, mammal, insect, plant, or the music of a mountain stream."[1] This, of course, is how Murie acquired his phenomenal knowledge of the natural world and, like the Kalahari hunter-gatherers, could recognize everything he saw and understand everything that was happening around him as clearly as the rest of us can understand a nature film with its predigested information carefully explained. Not only could Murie recognize the tracks and scats of every mammal in America, but he also knew the scats of different kinds of snakes, to say nothing of their tracks, and could tell the difference between the tracks of a bullfrog and a leopard frog or the scats of a sage grouse, a ruffed grouse, and a spruce grouse. Trust me, there's very little difference. He could even tell the difference between the tracks of a cricket (a Mormon cricket—he wouldn't just call it a cricket) and a grasshopper. Imagine the world that was open to Olaus Murie.

So I try to see more, to be a better observer—less like Liz Thomas and more like Olaus Murie. Sometimes I manage to do that, at least to a degree. Anyone can manage some kind of observation, and if you try, a cornucopia will open and one surprise after another will spill out at you, often as an unexpected by-product of some other effort. You just need to be ready.

But you will not be ready if you cling to the notion that our species is basically different from the rest. We're not, except that we're more destructive. One of the greatest barriers we have to understanding the life-forms around us is the burden of misinformation we carry in our heads. As it was with the residual farm lore that left me with many a misimpression, so we are often told that if humans have a certain characteristic, then by definition, an animal cannot. For a while, even certain scientists appeared to hold this view, or so it seemed when questions of anthropomorphism arose. In part because of the extreme caution of the scientists, fear of acknowledging shared characteristics has leaked out and is now in the general public to the degree that an editor at the *New York Review of Books*—surely a highly educated person—challenged my use of the term "naked" for an infant macaque monkey who was born, as are all macaques, with hair so fine and sparse as to be almost invisible. "Is it unusual for a rhesus monkey to be naked?" he asked sarcastically. "Do they usually wear clothes?" He seemed unaware that the dictionary allows my use of "naked," but what can one do? I withdrew the piece I had written. When my friend the writer Sy Montgomery referred to a female crane who had just laid an egg as "she," her editor struck out "she" and wrote "it." To attribute gender to a crane was anthropomorphizing. I've had similar difficulties with editors who, for instance, didn't like my use of the word *want* when speaking of what plants want, even though some human wants—those inspired by glandular information, perhaps—may be as hardwired as those of any plant. Nor did some editors like the use of "who" and "whom" for nonhuman species. I don't tolerate such ideology, and neither did Sy. I've used "who" and "whom" for trees. But these editors were only doing what they thought was right—victims of vastly misplaced political correctness that far too many people carry. So it's best to get away from that. A good way to look at other life-forms is to view them all as something like yourself.

Consider *Cordyceps,* for example, a simple fungus you might pass without noticing. When an insect crawls over it, it will attach itself to the insect's carapace and excrete a chemical to burn a hole. It then will enter the insect through the hole and begin to eat its nonvital organs. To protect the insect until the *Cordyceps* has finished with it, the *Cordyceps* will excrete an antibiotic against bacterial infection, also a fungicide against fungal infection, and also an insecticide against attack by other insects. After eating the nonvital organs, the *Cordyceps* will eat part of the insect's brain. Somehow, this will cause the insect to climb a tree and fasten itself to the top. When the insect has secured itself, *Cordyceps* will eat the rest of the brain. Then the insect will die, and the *Cordyceps* will split open the corpse and release its spores into the wind. What is all this for? The fungus wants its spores to travel far and wide, improving their chances of survival. This would not happen if they were released on the ground if the parent couldn't get up into the wind.[2] Thus the parent *Cordyceps* does its best by its offspring. We do that too, if not the same way. We are all the children of Gaia.

Thus a useful way to look at another life-form is to assume that whatever it may be doing—chewing bark, digging a tiny hole, wrapping itself in a leaf, sending up a sprout, turning its leaves to face the sunlight—it is trying to achieve a goal that you, in your way, would also want to achieve. In fact, you can be sure of that. The closer you are taxonomically to what you are looking at, the more likely you are to recognize what its goals might be, and the further you are, the less likely. Either way, it's fascinating.

For instance, I came home one afternoon in August and saw a group of four deer in the field across the road, but they were just entering the woods, and I couldn't identify them. If I had been there when they were in the field, I could at least have determined whether I knew them or not. I walked around for a while to see if I could learn more but I didn't. However, as I was leaving I saw an enormous green caterpillar, almost five inches long, coming out of the bushes on the east side of a path. He looked very much like a tomato hornworm, but I didn't think he was. His proportions seemed different and I knew of no tomato plants for miles around. But I am not an entomologist by any stretch of the imagination. I could only wonder and watch. Once across the path, he selected a very thin sapling about three feet tall and started to climb it, clasping the stalk with his pseudopodia, moving the rear one first, then the next and the next, finally grasping with his six true legs—the insect legs on his first segment—just as his rear pseudopod clasped the stalk for a second time. Up and up he climbed, all the way to the top. Once there, however, he found the plant not to his liking. He came down again and, continuing on his line of travel, he went in among the bushes to the west.

I was stunned. Almost exactly a year earlier I had seen exactly the same thing—the same kind of caterpillar crossing the same path at almost the same place, and climbing a small plant, perhaps even the same plant, only to abandon it in almost exactly the same way and keep traveling. I felt fairly sure that despite their astonishing similarities, these were two individuals, not the same caterpillar, although certain insects do remain in their larval state for years, including certain butterflies and moths. But I couldn't believe that many of them do this in New Hampshire, or not aboveground, especially if they are large and soft with naked skin. It's true that the little wooly bear caterpillars survive while frozen solid, but they are specially equipped for this with exceptional metabolisms. So it seemed to me that by the time I saw the second caterpillar, the first must have turned into a moth and surely was no longer living. But if that were so, what explained the almost precise repetition?

I had no idea, of course, although a few things seemed likely. I didn't think the caterpillars were looking for food because they would have been able to recognize the plant's species from afar. Also they passed by leaves on the way up and weren't interested in them. So perhaps they were looking for a place to pupate, chose a thin plant because they could grasp it with their pseudopodia, then found when they reached the top that it was too short and went to look elsewhere. At any rate it was a splendid example of a living creature, or two living creatures, struggling to meet their own needs, which surely were important. The second caterpillar—the one I saw after looking for deer information that August afternoon—had been in a hurry, perhaps almost in a desperate hurry, and had walked quickly. I couldn't remember if the first caterpillar, a year earlier, had been in a hurry too. But I felt sure, without a doubt, that something major had been happening in both their lives.

Most experiences of wildlife viewing are mixtures. Often you can understand a part of what you see, but not all of it. At the time of this writing my four-year-old grandson, Jasper, lives with his parents in a house across the road. He is most helpful in learning the surroundings, because every evening at his suggestion we go out to look for—as he puts it—"something interesting." His eyes are closer to the ground than mine, and his eyesight is better. He also looks up into many plants rather than down on them, and notes all manner of things that a taller passerby would never see. He points them out, and we try to interpret what we're looking at with variable success. One day we found a solitary milkweed surrounded by long grass and goldenrod. That in itself was unusual, as milkweeds normally occur in groves. Jasper knows many plants by name, and is much taken by milkweed with its milk and fluffy seeds, so he examined this one carefully and saw a monarch caterpillar on the underside of a leaf.

Oh wow! We were excited to learn what the caterpillar would do. But wildlife observation takes great patience. One must be prepared to wait. So we waited and waited, but the caterpillar didn't move. We asked each other if he was hiding, and also how he managed to stick to the leaf while upside down. We had no answers. To pass the time, we examined the milkweed more carefully. We saw a number of caterpillar droppings on one leaf, just one, and not the one to which he was clinging. What was this? We believed they were his, as they seemed fairly fresh and he was the only caterpillar present. Nor were other milkweeds nearby that could shelter other caterpillars who might have come over and done this. But we didn't think he could have made so many droppings all at once. I had seen a wooly bear make droppings, and if I remember rightly, he made only one or two. Here, there were at least ten or twelve—surely far too many to make all at once—so did this mean that our caterpillar visited that leaf when he needed to make a dropping? Many animals have latrines, but caterpillars? I knew very little about caterpillars, as did Jasper at that point in his life, and we had no explanation for the phenomenon. So we looked further and found tiny round holes in other leaves. We thought we knew what this meant—the caterpillar had been eating—but not very much from any one leaf, perhaps because the poison milk of milkweed tastes so horrible. Monarchs have no choice, of course—they eat only milkweed, which makes them poisonous, hence birds avoid them—but an injured milkweed can take a moment before it starts to ooze, so perhaps the caterpillar took his bites before the oozing started in earnest. Meanwhile we kept checking on him to see if he would move, but he didn't. It was getting dark. We decided to come back the next day.

And we did, again in the evening. In our absence the caterpillar had climbed the plant and was standing quietly in plain sight on top of a higher leaf. No droppings were visible this time, not even the ones from the day before. Had the wind blown them off the leaf? We didn't know. And as far as we could tell, no new holes had been eaten in the other leaves. Again we watched the caterpillar for a long time and again he did nothing. Could he have known that we were there and was trying to be inconspicuous? The thought was tempting, but monarch caterpillars are brightly colored for the purpose of being conspicuous. Anyone who eats one will never eat another. To be noticed should therefore not disturb a monarch caterpillar. Unless—and here another tempting thought occurred to us—he knew the difference between birds and other creatures, and might not have been concerned about birds but could have been concerned about us, with our giant shadows falling on him.

By then we were thoroughly engaged with this caterpillar and we decided to come back in the morning. But Jasper had a swimming lesson in the morning so I went back on my own. Yes! The caterpillar was still there, but on yet another leaf, and this time he was eating. If only Jasper could have seen it! The caterpillar was slowly taking tiny bites on either side of the leaf's central vein. I checked on him a few hours later and found that he had folded the leaf in half, like a hand palm up with fingers closed, and the fold was where he had been taking bites. Perhaps he had chewed into the central vein in order to weaken it and enable him to fold the leaf. Or if the vein was too hard for him to bite, perhaps he had removed the surrounding leaf tissue that would have reinforced it.

He was standing on top of what once was the underside of the leaf. Certain caterpillars roll themselves in leaves to pupate, but I've never heard that monarchs did this—as far as I knew, they find a suitable twig somewhere, spin a little silk that they grip with the rear pseudopod and hang upside down in the same position we would assume if we hung by our knees and bent our heads to look up at our chests. Their skin then loses its conspicuous striping, assumes a plain, leaflike color—not the color of a fresh, green leaf but of a faded, unappetizing leaf with tiny spots—and hardens into a smooth chrysalis from which, about ten days later, they emerge in butterfly form. That being so, I could think of no reason why this caterpillar had gone to so much effort to fold a leaf. But the act of folding probably wasn't random. Everything an insect does has purpose. As a friend once said, insects are never wrong.

I waited for Jasper to come home late in the afternoon. With great anticipation we went to the milkweed, but the caterpillar was gone. So was the folded leaf on which he had been standing. What could have happened? We searched in the leaf litter below the milkweed in hopes, perhaps, of finding him or at least the folded leaf, but we didn't. We never saw him again.

This does not necessarily mean that something bad happened to him, although the fate of the leaf was puzzling. A small maple tree was within his walking distance. Perhaps he went there to pupate. We looked at the tree, but it was in a large brushy thicket and we lost hope of finding something as small as a caterpillar hanging upside down.

Monarchs are amazing creatures, and have been carefully studied. They begin life as eggs the size of the dot on the *i* in *caterpillar.* When they hatch, they're the size of the *l* in *caterpillar.* They enlarge, of course, at first eating only the tiniest hairs of the milkweed, taking more substantial bites as they grow bigger. A caterpillar's skin is not like ours but is more like a flexible casing, not quite so much a part of the caterpillar as a container in which he grows. Meanwhile, a new skin is forming inside the casing, which soon becomes too tight, so the caterpillar gets rid of it, wriggling out of it and rubbing it off his face with his forefeet, just as we would use our hands. The final skin is the chrysalis in which he pupates. Our monarch was about two inches long, thus was probably about two weeks old, and would have shed his skin about five times since hatching.

His was the generation that would migrate, and thus he would live longer than the earlier generations of that summer, or he would if he survived the migration, which must be a devastating experience for creatures so light and so small. Now and then during the weeks that followed, small flocks of monarchs would appear on our flowers, where they would eat for a while and then depart—evidently migrants on their way to wherever it was they hoped to spend the winter. They would fly off to the south, their little bodies tipping this way and that as the light breeze took them.

Our prevailing wind is from the southwest, and would carry them toward the coast. Once there, they would need to follow the shoreline and not let the wind carry them out to sea. Some of them manage to do this—the Atlantic Coast is a flyway for monarchs—but I cannot picture how they manage with their tiny bodies that weigh about the same as milkweed seeds, and their little gossamer wings. We didn't know if our caterpillar was among those migrants, but he could have been. We watched television to see what the weather was doing, glad when the weather was warm and calm, worried when a storm came up the coast from Georgia. On September 28, as Hurricane Kyle was moving toward us, one lone monarch appeared and flew back and forth across my office window all afternoon. The window wasn't thirty feet from where we first saw our monarch caterpillar. Was it him, all alone, with a hurricane coming? We hoped it wasn't—we had last seen him around the end of August, and this was almost a month later. But it might have been him—he might have emerged from his chrysalis and stayed around. But wherever he was, we wished him well—him and all the other monarchs out there, struggling on with their astonishing, important lives.

If a goal of an organism is to live, some trees on my land were meeting the goal spectacularly. In 2008, Anne McBride of the Monadnock Conservancy and Brian Hotz of the Society for the Protection of New Hampshire Forests discovered—not a mile from my house on land that wasn't mine until 2003—a grove of rare, black gum trees between four hundred and six hundred years old. I began to visit that piece of land as soon as I owned it, but as usual, I must have been looking for animal signs and didn't take note of the trees. I was amazed and elated to learn of them. When Columbus set foot on our continent, those trees were saplings. When the first colonials began to clear what was later to become my patch of forest, those trees were three hundred years old. But here was a mystery. In the woods not far from where they stood were the ruins of an old mill and the stone foundation of a small house, and also miles of stone walls that once surrounded pastures. How had these trees survived in the presence of so much human activity focused solely upon farms and field making? I consulted some forestry books that belonged to my father, and in one of them—*American Forest Trees* by Henry A. Gibson, published in 1913—I learned that the wood is difficult to cut because the grain goes in different directions. When the area that became my land was first cleared, there were no chain saws. People cut trees with a two-man saw or, more likely, with an ax. But a dense, complex grain will buckle a saw and prevent an ax from dislodging a chip no matter where it strikes. In the book is the photo of a frustrated woodcutter who, with his ax, has hacked into a black gum dozens of times with no better result than dozens of ax marks. The tree remains rock-solid. Surely this explains the presence of these black gum trees. The old farmers didn't cut them down because they couldn't.

Gibson tells us that the foliage of these trees is spectacular in autumn. I have yet to see them in the autumn. Gibson also tells us that in October, they drop little fruits that no animals will eat. He doesn't say why, and doesn't pursue the question. But there is a question, a very important question, because the usual reason for a tree to make fruit is so someone will eat it, swallow the seeds, and carry them off to pack them in dung and drop them far away. If a tree is making fruit that no one seems to want, it suggests that the tree had a particular species in mind to do this service, and the species is no longer around. One thinks of the dodos, who lived on the island of Mauritius until the 1600s, when they famously became extinct. The tambalacoque trees of Mauritius are following the dodos into oblivion because, it is believed, the only creatures who would eat their fruit were dodos, and the seeds had to pass through the dodo's gut before they could germinate. The tambalacoque trees on Mauritius today have outlived the dodos by four hundred years, but they no longer reproduce, or not often, or sometimes only with complicated human assistance.

However, the black gum trees won't be needing human help, or, in fact, the help of any animal. Their groves contain trees of different ages, some younger than others, which means they are reproducing on their own. They send up sprouts from their roots or stumps, sprouts that become trees eventually. As far as I know, all trees can make sprouts, but most of them don't depend on sprouting for existence, certainly not in New Hampshire. Maple trees have winged seeds that the wind carries, oaks have acorns that the squirrels plant, wild cherries have fruit that birds eat, and so on. But according to the New Hampshire Natural Heritage Inventory of April 2000, it was the tendency of black gum trees to produce sprouts (plus their tolerance of hundreds of years of hurricanes and other disasters) that explains why relatively young black gum trees are found amid the truly ancient ones. And because a sprout rising from a stump or root is not the offspring of the original tree but an extension of it, a black gum that today is six hundred years old may have sprouted from a stump that was already six hundred years old, hence the total age of that particular tree could be 1,200 years or older.[3] That being so, who knows for whom the fruit was first intended? Someone must have eaten it, but whoever it was must have vanished in the very deep past. And still, the black gum trees keep fruiting.

The older trees are tall, of course, but seem surprisingly thin, with straight, spiky branches and deeply furrowed bark. One of the oldest black gum trees in my forest has a large hole in the trunk about ten feet up, the home of a fortunate animal. The botanical name of their species is *Nyssa sylvatica,* or nymph of the forest. Few things could be less like nymphs than these tall, ancient beings with their heavy, rough bark. Nymphs of the forest? Who thought of that?

Then there were the oak trees. Those, I thought I knew. Ten of them were within thirty feet of my house and, unlike the black gums, they were only in their eighties or nineties. They appear in a portrait of my mother, painted in 1937 by her closest friend, the artist Grace Reasoner Clark, who posed my mother outdoors with these trees behind her. My mother was thirty-nine, I was six, and the trees were graceful little saplings. Today, they are ninety feet tall and sixty inches around at chest height, the offspring of a massive parent that once stood in the midst of where they are now, but, unlike the black gums in the forest, did not survive the hurricane of 1938. My grandmother and I were looking out the window and saw this oak tree fall. First the wind pulled off its leaves, then made it lean sideways, then pushed it over with a tremendous crash that vibrated the house. The tree was still alive but doomed, lying on its side with its ruined branches on the ground, their few remaining leaves still trembling, and the giant disc of its roots still holding earth, exposed and gaping. After the hurricane passed by, my grandmother and I went to look at it with pity. If we could have stood it up again and saved it, we would have.

The offspring of that tree grow close together along a stone wall, where the parent shed its acorns. The largest one is on the east edge of the grove where it gets the most sunlight. A maple grows right beside it, an intruder that to some extent blocks the sun. So the oaks compete with one another and with the maple. Trying to live as well as they can, they extend a branch or two far out over the driveway where sunlight can reach the leaves. The largest, eastern tree does that the most. I hate to say this, but over the years we have sometimes had to cut the branches because they scrape the tops of delivery trucks and the drivers complain. The tree will then promote another branch, encouraging it with lots of nourishment, and in a few years, that branch will also be over the driveway. Like the hand of a supplicant, it will reach for the sun until it also becomes large and heavy. It then begins to sag, scrapes the top of a delivery truck, and we cut it, but we aren't happy about it.

These trees withheld acorns in 2007. I somewhat resented that, although they were only looking out for themselves, just as they should, but otherwise I took them for granted and seldom thought about them except to wonder if, in a strong wind, one of them might fall on the house. However, in the autumn of

2008 I began to take more notice because, in contrast to the previous year, the trees were developing acorns. They would feed the deer and the wild turkeys. Unlike the black gum trees with their unpopular fruit, the oaks fed almost everybody, including, of course, the squirrels who, in contrast to the other acorn predators, were helpful and planted some of the acorns for them. A squirrel will bury many acorns to eat later, but will not find them all. These will grow into trees and one day will produce more acorns that will feed more squirrels. That's the oak tree's reproductive strategy. It's not as fancy as *Cordyceps*, but it works.

One day, the acorns began to fall. We heard them pounding on the roof. Interestingly, we would hear all kinds of clattering for a while, then as much as an hour of silence, then suddenly much more clattering. I hadn't noticed that before and I went to examine the oaks more carefully. Up in the crown of the largest, easternmost tree I spotted a cluster of six or seven acorns. Then suddenly I heard a pop and saw them burst away from their footings all at once. They didn't just drop—they flew off, as if in response to pressure. Soon another cluster popped and fell. Within moments, the next tree to the west began to do this, and then all were doing it. Acorns rattled on the roof of our house like machine-gun fire. And just as suddenly, silence.

What had I seen? The trees beside which I had

lived for much of my life were doing something I had never noticed. I couldn't give myself high marks for observation, but at least I thought to notice if the wind was blowing. It wasn't. The trees were doing this all by themselves. Since I am always looking for intent in life-forms, asking myself what does this plant or animal want (or, to seem a bit more scientific, what is the purpose of what it's doing), it came to me that the trees might be coordinating. They were, after all, siblings. And trees do coordinate—the year before, the oaks had coordinated to withhold acorns, and I've heard that if insects attack a certain kind of tree in a grove, the tree releases a pheromone into the wind to warn the other trees who then produce a toxin and are ready for the insects when they come.

What could be more gripping? Were the oaks really coordinating to fling out their acorns, or was I imagining? I kept watch and came to the conclusion that while the coordination could possibly be caused by an unknown environmental factor, the acorns really did fly off. It made sense. If the acorns dropped straight down, they might land on the thickest part of their parent's roots, where they would have trouble inserting their own roots. Better to land a few feet away. But the most interesting thing of all—although I can hardly believe this myself because it negates the possibility of the environmental factor (and I won't be surprised if others can't believe it either)—is that the biggest oak to the east seemed to initiate the action. Was this the oldest sibling, showing the others how it's done? Not likely. Even so, whenever I was there to watch this, the tree to the east went first.

I made a feeble effort to learn more about acorns, but didn't. I did learn something about leaves, however. I learned that oaks originated in the tropics, which is why their bark is relatively smooth, like that of tropical trees. When the world's landmasses readjusted themselves, the oaks rode their continent northward and found themselves in a different climate. In the tropics, they would not have shed their leaves, and this is why, even today, they are among the last trees to lose dry leaves in the fall. Beech trees lose their leaves even later, and like the oaks, they too originated in the tropics. The last leaves don't fall off until spring, when new buds push them.

The phenomenon did not go unnoticed. A man who made a deal with the Devil knew about it, and when he promised his soul in exchange for a favor, he told the Devil he could have the soul as soon as the oak trees were bare. The Devil agreed but was outsmarted, because the oaks are never bare. Always, a few leaves are clinging.

Insects and plants can leave you filled with wonder, but your questions can't always be resolved. If a fungus wants what we want, if a tree wants what we want, and if insects want what we want, how much more obvious could it be that a vertebrate wants what we want? This brings us to the mouse and rat families. Plenty of them live in my house or in the woods and fields—voles, wood rats, and white-footed mice who are indigenous, also house mice who are not indigenous, originating as they did with brown rats in the lands that reach northward from Iraq to the shores of the Caspian Sea. They, with our cats and the bread we eat, represent a little ecosystem held over from the days when people domesticated grass. Eight thousand years ago, as the glaciers were receding from New Hampshire, Neolithic people between the Tigris and Euphrates rivers began to collect the seeds of the wild grasses that were to become wheat and millet. They stored the seeds in their granaries. The local rats and mice—the familiar creatures now known as brown rats and house mice—had been eating these seeds all along, and, glad to find them in large collections, also went into the granaries. Right behind them came their predator, *Felis sylvestris lybica,* the little tabby wildcat from whom domestic cats descend. In time, this closely tied group began to move to other places, wherever ships carried cargoes of grain. When grain was being loaded, some mice and rats might get on board with it, a few observant cats might follow, and away they would sail to populate the planet. Now the little ecosystem of wheat, mice, and cats, if not always rats, is part of almost every household in the so-called civilized world. It's part of mine, and while these Near Eastern species may not be indigenous, they're here now, and I value them. I too belong to an invasive species that began on the African savannah, the only primate ever to reach our part of the Holarctic, and when it occurs to me how foreign I am, surrounded on all sides by those who evolved here and are equipped to live outdoors year-round, I sometimes feel isolated.

But I needn't—not if mice and rats are near, as these are our closest relatives in North America. We have the same ancestor. And although that ancestor lived during the Cretaceous, some of our similarities have firmly endured. That's why mice are used in laboratory experiments as substitutes for human beings. That's why the federal government has sponsored the "mouse knockout" project. Scientists are knocking out each of the twenty-thousand-odd genes in the mice genome one by one. Doing so will tell us a lot about the functions of our own twenty-thousand-odd genes, which closely match the mouse's. And there's really no surprise that we have genetic similarities. We have behavioral similarities too—strong ones. For instance, we are successful breeders and have large populations. Also, unlike many species such as the deer, our families include adults of both genders. We also make sure we have supplies of food against the time when food might not be available. We are also good students, learning quickly and easily. As an animal trainer once put it, "You never have to tell a rat a thing twice." Thus we can solve complex puzzles. We live in houses—rats and mice make their own homes if they don't use ours. We have hands with which we manipulate objects and wash our faces and behind our ears. We don't like to be controlled by others—we are happy if we manage our own lives, as was demonstrated by a white-footed mouse in a laboratory. A scientist wanted to know if this mouse's species preferred the soft light of dawn and dusk to full sunlight. So the mouse was placed in a box with two dimmer switches, one inside for the mouse to use, and the other outside for the scientist. But the experiment didn't work as planned, because the mouse didn't care about the light—he cared that he was being manipulated. When the scientist turned up the lights, the mouse turned them down. When the scientist turned them down, the mouse turned them up. He quickly saw that something was purposely changing his environment and he wanted to take control. He must have felt as we do when a small child snaps a light switch off and on. It makes us crazy, and we immediately put a stop to it. Thus the unplanned result of the experiment was not surprising, and perhaps had its roots in the Cretaceous, where, like many other animals today, our ancestors must have valued their autonomy. The most surprising similarity we share with mice is that we sing, not just squeaks or howls, but actual music, and both our species do it well. The least surprising similarity is that we eat the same foods.

Mice can sing? They can and they do. One day, before my mother came to live with me, I visited her in her house in Cambridge. My father was no longer living, and the big empty house was quiet—only my mother and her elderly orange cat were in it, and the mice who lived in the basement. The cat had given up mousing because he had lost some of his teeth, but even so, the mice were seldom in evidence. Like all such mice, they would have had a colony with thirty or forty members gathered into family groups of several grown females and their children, a few young males with little or no authority, and a dominant male with much authority. If mice from other colonies were to appear, the mice of my mother's colony would massacre them or drive them off. Thus it seems safe to say that my mother's environment, friendly to animals and very stable, would have had just one regular, functional colony of mice numbering somewhere around forty or even fifty. Now and then we'd see one run across the floor. My mother might put down a little piece of cheese. The mouse would come back in the dark and take it.

On this particular day, she had opened the front door to let me in, and we were standing together in her hallway. Just then, we heard birdsong, or thought we did. It was coming from her kitchen. A bird in the house, not flying in panic against the windows but singing? We went to the kitchen, and there we saw a mouse sitting upright at the edge of the counter with his front paws crooked against his chest, singing one of the most beautiful songs that we had ever heard. I had no idea that mice could sing, and for as long as I live, I will never forget it. For a moment he didn't see us and kept on singing. Then he saw us, jumped down to the floor, and ran.

I am fascinated with mice for many reasons, but especially because they sing. They sing for the same reason that wolves and birds do, a reminder to others not to trespass, and perhaps for other reasons too. But they don't sing often, or not in my experience. I've heard from other people of other mice singing, but in a lifetime of living in houses with mice in them, I heard only that one song.

As for two-parent families, I once caught a pair of house mice and put them in a big aquarium. They soon had six children. It is easy enough to identify gender in adult mice, so I was surprised to see the father taking care of the infants while the mother occupied herself nearby. I hadn't expected that. As soon as the infants grew a little fur, they began to move around, but this worried their father, and he would pile them all together in a heap and spread himself on top of them, covering them all as best he could, with his arms and legs stretched over their bodies. Why? It doesn't take much to overcool an infant mouse, so most likely he was trying to keep them warm. If one of them managed to wiggle away, he would quickly drag it back to the pile of its siblings and spread himself over them again. When the infants were old enough I tipped the aquarium gently on its side to give the group the choice of staying or leaving, and they left. I expect they rejoined their colony, wherever it was. I was a bit surprised, however, that they never returned to the aquarium for food, although I left it on its side with food in it. But they never came back to the scene of their captivity, not even for peanuts and cheese.

We have always had mice in our house and in my office—two kinds, the gray, foreign, house mice and the red-brown, indigenous white-footed mice—so it would seem we had more than one colony. However, they all stayed mainly out of sight, or they did while a certain cat was in residence. His name was Pula and he was a champion mouser. Unfortunately, he was also an occasional birder, although not in the league of the great James Audubon, whose collection of specimens enabled his art. And as far as I'm concerned, the cat had the high ground because he ate his victims while Audubon just drew pictures of them. Anyway, when Pula wasn't mousing, he was traveling, and sometimes he didn't come home before dark. Then I'd go looking for him in the car, and if I saw him by the roadside, as I often did, I'd pick him up and bring him back with me. I thought he appreciated this. I thought he was glad to be retrieved and didn't have to walk home alone when the coyotes and the fishers were out hunting. But now I'm not sure. One day, to learn what he was doing when he wasn't with us, I fitted him with a little radio collar. To my sorrow, I found he was visiting other families up and down the road in search of a home that was more to his liking. Perhaps because of personality clashes with our other cats, he had decided to move on.

At last, after numerous visits and much consideration, he chose a family about a mile away. Their elderly cat was no longer living, so they had no cats, which may be why Pula favored them. They liked Pula and took him in. For many years after that, I'd see him at the roadside when I drove by at night, and I'd stop and get out of the car. He always recognized me, and he'd hurry over for a visit, purring, but if I tried to pick him up he'd make an almost inaudible growl, just a suggestion, to tell me

I should put him down and not try any more to take him home with me. Like the mouse with the dimmer switch, he wanted to control his own life. At the time of this writing, I still see him occasionally by the road at night and we have our little visit.

As has been said, while Pula lived among us, the mice stayed in the walls. If they were unwise enough to come out, he caught them. Our other cats were getting on in years and didn't bother very much about the mice, so after Pula moved away, we began to see mice often and found their droppings in the desk in my office and in the kitchen drawers. Without Pula to help me, I had to clean the drawers almost every day. I set traps, but those particular mice were so accomplished that they could eat the bait without springing the trap. Of all the mice that lived in our house, I managed to trap just one. So when Pula and I would stand together in the dark by the roadside while light streamed from the windows of his new home, I would feel a twinge of envy, knowing that those people were mouse-free and we weren't, and all because this cat knew his own mind. He continued to like me well enough—he always welcomed me when I'd see him at the roadside—and because his new family let him come and go as he pleased, he could easily have come back to our house if he wished. I hoped he would. But he did not. His decision had been final.

After that, thanks to me and also, perhaps, to Pula's absence, as he might possibly have been able to help with the situation, we had a truly major tragedy. What led up to it was an event involving my parrots. They lived in my office in giant cages with bars about an inch apart, so when, one night, a raccoon came in through the dog door, he couldn't touch the birds, but he could reach into their dishes for their uneaten food. Something about him so panicked my macaw that she went into hysterics, beating her wings and screaming so loudly that I heard her from the house and came running. The raccoon was gone but the macaw was inconsolable. She had one panic attack after another, wildly hysterical and trying to bite, and it was me, my husband, and the dogs who set her off. Although she had shared the one-room office with the dogs and me ever since her adolescence, and although I unfailingly cared for her every day, she seemed not to know any of us. I still had to feed her, of course, and give her water, and luckily the cage had little doors for the dishes so I didn't have to get inside with her. I was afraid that if I did, either she would bite me so badly that I'd lose an eye or a finger, or she would literally be frightened to death.

During the second day of the macaw's intermittent hysterics, my African grey parrot flew to the top of her cage. She didn't like anything on top of her cage and didn't particularly like him but on that occasion, she more or less ignored him. Thinking that she was perhaps okay by then, I tried again to approach her. But again she went into hysterics. Not until the third day was she calm enough to let me near her, and by then I thought I knew what this was all about. My macaw understood that the people, the dogs, and the raccoon were the same kind of creature. All of us were mammals, thus more like one another than we were like her. Nor were we like the other parrot, who wasn't the same species as she was but was nevertheless a bird. And I wasn't. I belonged to a Class of dangerous animals and in her eyes was a kind of raccoon.

I was deeply impressed. I've met kids in the local high school who couldn't reach that level of taxonomy, or not without a lot of prompting. The people who think that animals lack cognition should revisit their views.

The episode with the raccoon played a brief but important role in the tragedy I generated and for which I cannot forgive myself, when a colony of rats moved into the wood storage shed that adjoins my office. It's in a building that was the barn of the original farmhouse when my father bought the land, and although the old barn was reconstructed, it kept some of its original populations. It's attic was once a refugium for flies, for instance, many thousands of whom would shelter inside it in winter. As usual, I didn't know just what kind of flies they were, but they looked like houseflies. They'd be pretty much dormant, of course, and never bothered anybody, and in the spring, they would leave. All summer we'd know nothing about them—we didn't have flies in the house, or not many, and nothing much was lying around outside that would attract them, as our garbage was contained and the dog droppings were disposed of, but as soon as the weather got cold, the flies would be back in their multitudes, knowing that winter was coming and they needed shelter to live. They'd been using the attic for a very long time—they were there in their numbers when I was a child, and probably long before that. The refugia of houseflies are ancestral. They are not abandoned unless something bad happens. And it did. My dogs got fleas, and I decided to fumigate the office, and though I killed the fleas, I killed the flies too, and never again was my office a refugium. I would have welcomed flies if more came back, but though there were plenty of flies in New Hampshire, perhaps using other barns as winter refugia, none were still living who knew about mine. I had wiped out a coordinated population. I didn't mean to—I was just thoughtless—and it was tragic. But it wasn't as bad as what happened after that.

I first learned of the rats from their droppings, which I found on the floor near the parrots' cages. I was very busy and didn't think much of it—not being Olaus Murie, I didn't know the difference between the droppings of a wood rat and those of a brown rat and wouldn't have worried if I did. I was always finding something like those droppings because my office is porous and often enough small creatures come in through the various cracks and holes. A few days later I was working late. My dogs were lying beside me when one of them suddenly leaped to her feet and ran across the room, and when I looked, I saw a small rat disappear in the crack behind the chimney. A brown rat! So that was the kind of rat that left the droppings, and there wasn't just one—the droppings were too many. Obviously, there were quite a few rats, perhaps a small colony. I then began to worry about the parrots, especially the macaw with her phobia about mammals. I wished for Pula. If he were with us, he would catch some rats and frighten off the others. But he had lent his hunting skills to others far away.

So, although my other cats were not as dedicated as Pula, they were the only cats we had, and I decided to leave them in my office for the night. But in the morning, I couldn't find them. Worried, I called them, but they didn't answer and didn't come. At last I found them squeezed behind the books on the upper bookshelves, hiding, and even when they saw me they didn't want to come out. I realized they had been frightened out of their minds. Perhaps the rats were worse than I thought.

The following night, I put the dogs in my office. I had a small cot there, and in the morning I found the dogs together on it, somewhat subdued, their ears slightly folded, their eyes almost accusing. They quietly got down and came to greet me, then silently went past me out the door in the manner of people who are politely trying to bypass an unpleasant situation. Obviously, we had a problem. I had brought the parrots their breakfast, and when I came to the macaw, she again became hysterical. She shrieked and threatened to bite. *Stay away, you horrid thing,* her manner said.

If not for the animals of our family—the dogs, cats, and parrots who have no choice but to live with us—I would have let the rats come and go as they pleased. I would even have put out a little food for them. They would have gotten used to me and gone on about their business in my presence. I could have learned as much from them and about them as I was later to learn from the deer, if not more. But as things were, my first obligation was to my own animals, so the situation with the rats could not continue. Something had to be done. I closed the dog door and blocked all the holes I could find through which a rat might enter. In the morning, there were more droppings, some almost the size of kidney beans. So that didn't work. Later that day I saw who made the larger droppings. I was working at my desk with the dogs nearby, when suddenly the skin prickled at the back of my neck and I looked up, saw my dogs sitting stiffly with their eyes wide, their ears high, and their fur bristling, and then saw, on the far side of the room, an enormous rat about eighteen inches long, tail included, who must have weighed upwards of one pound. I took him to be the alpha rat, the silverback of his group, and I began to understand the extent of the problem. Being very much smaller than most of their many predators and also fearless, rats have found the surprise attack to be useful, and if necessary they will fling themselves ferociously at their enemies before the enemies can collect their wits and respond. Perhaps something like this had happened with the dogs. They appeared to respect this rat. They certainly did not go after him as they had gone after the bear. The rat looked at me for a moment, perfectly calmly, then went behind the chimney. It would seem that the rats had reopened that particular hole.

I told various people about the rats, and got all kinds of advice and warnings. The rats would destroy my books and chew through electric wires and set the building on fire. I had to get rid of them but wasn't sure how. In the past, I had tried to discourage mice in my mother's apartment by putting coyote urine (available commercially and used to keep deer from eating ornamental shrubs) behind her washing machine. It was a bad mistake. I don't know if it discouraged the mice, but it overwhelmed us—the vapor from the urine was so strong it stung our eyes, and although I quickly did my best to clean it up we couldn't use the apartment for a week.

I planned to get a Havahart trap. But my advisers told me what I already knew, the rats would probably come back. I hated to do it, but I set a few conventional traps—the kind that crush the victim's skull or break its neck—and these caught two of the smaller rats, but then no others, although the traps were sprung and the bait gone. As a last resort I put out poison. It comes in a small tray that you put on the floor. Mice and rats think it's food and carry it to their nests, where they eat some and store the rest against a time when other food may not be available. In the morning, half the poison was gone.

The next morning, the tray was empty. The next day I found a dead rat in my office. The day after that I found a large weasel, his mouth open in a grimace, his body contorted, lying dead on the lawn in front of my office. The day after that our neighbor Don found an enormous rat dead in the driveway. Not knowing why the rat had died, he threw the corpse in the field. When he told me about it I looked for the corpse but didn't find it. Perhaps a coyote had found it first and had himself a meal. The rats were dead, the mice too, and the poison was spreading. I realized what I had done and tried to find where the rats had stored the rest of the poison, but I couldn't. I tried to find more corpses, and I couldn't. Evidently they too had been eaten. This meant that in the woods, there would soon be other corpses, but there was nothing I could do about that except cry.

Truly, this was the worst thing that I have ever done, a serious crime, a multiple murder. Killing the flies was bad enough. This time I had killed my own kind—other mammals. Mindlessly, I had done to some of the co-owners of my land what serial killers do to people, and in a very cruel way at that. Judging from the appearance of the weasel, death by poison is nothing like a hunter's bullet in the brain—it is horribly painful to the degree that those who are affected don't hide, as do most animals in extremis, but go out in the open to die.

I would prefer not to say what I had done, rather than announce it in print. I was tempted to leave it out of the book, or to invent a happier story. The only reason I didn't, other than that the book is nonfiction, is to tell whoever will listen how dangerous and destructive poison is, so that others may refrain from using it. It's one thing if you live in the city and don't mind causing horrible agony to animals. The animals you murder will probably not be eaten by other animals unless they are found by someone's dog or cat. It's another thing if you live in the country, because you will do what I did and spread this terrible death widely. And you'll never know where it all ends. The weasels will eat the corpses of the rats, the coyotes will eat the corpses of the weasels, and who knows who will eat the corpses of the coyotes, or how long the death-food will last in their bodies. It could be a very long time.

You cannot make up for the evil things you do—they're there forever. You can only add better things to your list of deeds in hopes of creating some kind of cosmic balance. A few years after this disaster I put an easement on my land to preserve it, hopefully forever, for the plants and the wildlife, or in other words, for the co-owners of my land who survive me.

I would like to say I placed the easement in atonement, but in fact, our family had been planning the easement for many years. It was a lengthy process and very expensive—the land must be surveyed and appraised, legal advice is required, legal documents must be filed, and the entity that takes the easement—in my case the Monadnock Conservancy with the Society for the Protection of New Hampshire Forests as a backup in case the Monadnock Conservancy somehow disappears—requires a fund for supervising the land in perpetuity to make sure that whoever owns it is keeping the rules and not destroying the forest or building developments. The citizens of Peterborough, the town I live in, and of Greenfield, the next town over, where my land extends, have voted to create funds to help people such as myself with the expenses, which was fortunate. My husband and I could not afford to do this on our own. These funds are the best possible tools for conservation—to purchase large tracts of conservation land costs millions of dollars, while to help a landowner put an easement on the same amount of land costs only a few thousand dollars. The result is the same but with important differences. Land owned by a town generates no tax money, and taxpayers must support it if expenses arise. In contrast, land under easement pays taxes as before, and the landowner pays the expenses.

When my land is added to the Wapack Wildlife Reserve to the east, land that was donated by my father, about two thousand acres will be preserved. Adjoining these acres to north and south are other large conserved areas contributing to a corridor of conservation from Temple Mountain in the south, which was heroically saved by John and Connie Kieley, to Crochet Mountain in the north, where the Crochet Mountain

Rehabilitation Center maintains a haven of undisturbed forest. Wonderful as this is, much of the land is upland, and many life-forms need lowland with water. Mine is lowland, and with any luck, the lowland will remain in good condition after I am gone. The deer will rut in the fall and have their fawns in the spring, a bobcat will do the best he can to keep other bobcats off his territory, and the black gum trees will drink from their wetland, raise up their leaves to catch the sun, and live another century or longer.

Someday I'll be part of it too. As its custodian for all these years, I'd like to stay with it. And when my husband's ashes are mixed with mine and with those of our dogs and are scattered at the edge of the field where the deer come out, we will be part of it. The rain will wash us into the soil and we will have eternal life, not of the conscious mind, I'm sure, but definitely of the body. My father told me so at the time he told me about blood and chlorophyll. Matter is neither created nor destroyed, he said. In the brook in the woods, beside which we were standing when we had this conversation, atoms that once were in the dinosaurs were floating by, he told me. So we'll become part of the oaks, the mice, the deer, and the bobcats. We won't be part of the black gums, sadly, because we'll be on the wrong side of the road. But no matter where we are, or which life-forms our atoms may join, we will still go on, not in some cosmic afterlife, but as a regular part of our planet. Perhaps our planet isn't much in the general scheme of things—just a mote of interstellar dust on the far edge of the Milky Way, circling a tiny star that must be smaller than a pinprick in the eyes of god—but it's our mother.

Epilogue

On the first of October, soon after we mowed the field, the Deltas stepped out of the woods. First came the mother, then the two grown daughters, then the fawn, all gray in their winter hair. In their calm, everyday manner, they spaced themselves at a normal deer-distance and ate a little grass. Then they walked back to the woods. The once frantic young doe, the smaller daughter, was reunited with her family, and the fawn, for all the doubts surrounding his existence, was large and healthy, as ready to start a winter as a fawn can be, as if there had been no questions as to his well-being. Obviously, from the point of view of the Deltas, there had been no questions. I have seldom seen anything more normal than the brief appearance of these deer.

A few nights later as I came up the driveway, my headlights shone into many pairs of eyes—many deer, far more than three. I quickly tried to count them, but the deer were moving around and not all of them were looking at me, so I couldn't be sure. But at last I recognized them. They were the Deltas and also the Betas, back from wherever they had spent the summer. I couldn't tell if all the Betas were present, or if there was a fawn, but I knew it was them, right there where they belonged. Soon after that I noticed a few deer droppings near our oak trees. I would have liked to look more closely at the droppings, but a dog ran past and ate them. However, their presence meant that the deer knew about the acorns, and knew there were plenty.

This raised an important issue. I had put out corn the year before because the acorns had failed so completely. An abundance of acorns raised the question of whether or not I should put corn out again. I experienced a twinge of conscience. In addition to all the discouraging information put out by Fish and Game as to why people should not feed deer, they had yet another reason, again in the *New Hampshire Hunting Digest*. It is as follows: "Over the years, those who feed deer begin to feel as though the deer they are feeding are 'their deer.'"

This is true, and it has consequences. The deer-feeders become protective and post their land, and the more land that is posted, the fewer places there are to hunt and the fewer legitimate hunters there are to support the New Hampshire Fish and Game Department, already underfinanced. One hates to think what would happen to the deer and indeed, to all wildlife, without this Department, without the game wardens, without the field biologists.

If someone from Fish and Game were to ask me who owned the deer on my property, I'd say they belonged to the state. This would be the correct answer, but not a completely honest one, because in my heart the deer do seem to be somewhat mine, and not only them, but all the local deer who need a little help in winter. For all I know, they might need corn even if acorns are available. I think of the magnificent, ten-point buck whose corpse was found on neighboring land after he had been shot by a trespassing poacher. He was no longer contributing his superior genetic material. Young deer fathered by lesser bucks might lack his survival ability. Some of them might need assistance.

However, I know that Fish and Game has a point about deer-feeding. I kept tossing the issue around in my mind until November 20, 2008, when something happened that erased my doubts.

The autumn had been mild. But before dawn that day, the temperature plummeted—down to 11°F with windchill, the coldest day of the year by far and a record low for that time in November. As the sun went down, the air grew even colder and the wind picked up. On the short walk from the house to my office late that afternoon, I felt my fingers freezing. Even my office was colder than usual, but I went back to work anyway. Just then the flock of turkeys at the edge of the woods began to walk across the field. Night was coming, and they seemed to be heading for the place near the pond where they usually gathered in their pear-shaped flock before flying one by one into the trees. But instead of going there directly, they took a detour that brought them in front of my office. There, they stopped, facing my window. They could see me through the glass. I realized that their purpose was to look at me. They did this for a while, then went to poke around at the places where during the past winter I had put their corn.

Then the Deltas stepped out of the woods—the mother and both her grown daughters. This time the fawn was not with them. The rut had started—perhaps he was in the woods with other males. His mother and sisters nosed around in the grass for a minute. That seemed normal enough, but then they too came toward my office, stopped about thirty feet away, and also looked at me through the window. They looked and looked, as if searching my face. Then the mother turned and went to investigate the places where I had put corn. Her daughters followed. The turkeys were by then on their way towards the pond, but when they noticed the deer at the corn area, they came back, perhaps to see if they had been mistaken and some corn was there after all. There wasn't, so all together, the three deer with the turkeys surrounding their legs like three people walking amid a pack of hounds, they went off toward the woods.

I hadn't thought to put out corn so early. It was hunting season, and I didn't want to lure deer into the open at that time. Besides, the acorns and the other natural foods that were still present in the woods and fields were surely more nutritious than corn, and better for the deer and turkeys. But evidently these awesome creatures knew who had fed them the previous winter and wondered if she remembered them.

Then, at the edge of the woods, two more deer appeared. One was tall, very dark, and in beautiful condition. The other was smaller. They looked toward my office but didn't come forward, and instead began to graze. But every now and then they would look in my direction, so I got my field glasses and to my amazement, I saw they were the Taus, the mother and one daughter who looked like one of last year's twins. I hadn't seen these deer since March. Although three members of this family had been missing since winter and perhaps were long dead of hypothermia and starvation, here were the survivors, possibly now as residents, on the day that their nonwinter vanished and winter as they knew it reappeared.

The philosopher Ludwig Wittgenstein has famously said that if a lion could talk we wouldn't understand him, meaning that the minds of animals are beyond our reach. This was probably true for Wittgenstein, but it isn't true for me. As far as I'm concerned, compared to many other life-forms, deer and people are practically the same thing. Wouldn't the approach of winter, as expressed by the sudden, dramatic drop in temperature, rather than by shorter days or some other factor, remind these animals of what they were soon to experience, of what hunger and snow and biting wind would soon be doing to them, and make them think of life-giving corn? No deer had been in the field the day before, or the day before that, nor did they appear for days thereafter. Nor had they ever approached my office, not as they did on November 20, to search my face for information. And as for the turkeys, except for that one visit when they came to look at me, they had spent the summer and fall in or near the woods. But all of them came on that first day of intense cold—all but the Delta fawn, who didn't yet know about winter or hunger or corn—and all came at the same time, as they had at the start of the previous winter. I knew why they came, without a doubt. The following day I bought two hundred pounds of what they were thinking about, the first such purchase of the season.

If the Old Way still prevailed, as it did for hundreds of millions of years without our meddling, things would level out eventually and I wouldn't need to worry about the deer. But we meddled. And we continue to meddle constantly, in almost every way imaginable. However, we do get to choose how we meddle. I don't believe that the deer are my property, any more than other people are my property. Only my dogs are my property, but dogs are slaves, god bless them. The deer belong to themselves, or they belong to Gaia, and in the winter of 2007–2008 I chose to meddle with them because they were starving. But Fish and Game makes a good point in saying that people who feed deer soon wish to protect them. That obviously happened to me, and every winter from now on, for as long as I can carry a bucket of corn, I will try to protect the deer who live where I live, not because I think they are mine but because I know who they are. My husband will do this for me when I'm no longer living, and when our ashes are scattered at the edge of the field, maybe Jasper will do it.

I am indebted to the deer. I am also deeply indebted to the turkeys, but I managed to learn more from the deer. I learned how they protect themselves and how they feel about one another. I learned that they, far better than ourselves, understand the power of winter. And I learned that as I own my piece of land, so do many other life-forms, including deer. Our ownership is secured by different laws, but it's the same land. As the deer know me with deer knowledge, so I know them with human knowledge. To gain that knowledge I broke some rules, but these were rules of human invention. By the rules of the deer, I probably did quite well. As it was with the alliance of chital deer and langur monkeys in Central India—the langurs who dropped leaves and the chitals who ate them—we benefit from each other. The deer get food and the primate gets information, also recognition, to say nothing of the joy of knowing they're alive.

Notes

Chapter Two: Cracking the Code

Bauer, Erwin A. *Whitetails.* Stillwater, Minn.: Voyageur Press, 1993, 59. We know nothing of the kind. It's been shown definitively that the ability to reason is not related to brain size—witness Alex, Irene Pepperberg's famous parrot who, with his walnut-sized brain, stunned the world with his cognitive abilities.

Chapter Three: Deer Families

Rue, Leonard Lee III. *The Deer of North America.* New York: Lyons and Burford, 1997, 326.

Chapter Four: The Hazards of Feeding

[1] Nelson, Richard. *Heart and Blood: Living with Deer in America.* New York: Knopf, 1997, 73.

[2]*The Deer of North America,* p. 387.

[3]Ibid., 461.

Chapter Five: Deer Seasons, Human Seasons

[1] All the other primates have hair except for newborn infants. So why are we the only ones who are naked? Some say that we are infantilized versions of our forebears, just as most domestic animals are infantilized versions of their wild ancestors. However, there may be a much better explanation. Of the 233 species of primates, only two kinds—ourselves and the baboon types—live outside the forest. When we were forced by climate change to become savannah animals we had to adapt to the broiling sun. We hunted as our primate ancestors hunted, by the chase and grab method, and as we became proficient hind-leg runners we were able to capture larger game by running the animals to exhaustion. To do this we had to withstand heat better than the game we were pursuing, and hairlessness helped us to do so. The Kalahari hunter-gatherers, most especially the Ju/wa Bushmen, hunted by running for as long as they kept the old lifestyle, even though they hunted mainly with bows and arrows.

[2] *The Deer of North America*, 332.

[3] *Heart and Blood*, 32.

[4] Rue, Leonard Lee III. *Way of the Whitetail.* Stillwater, Minn.: Voyageur Press, 2000, 98.

Chapter Six: Fawns

The description is from *The Deer of North America*, 257.

Chapter Seven: Drivers, Hunters, and Their Prey

The Deer of North America, 341.

[2]Ibid., 342.

[3]Ibid., 342.

[4]Ibid., 343.

Chapter Eight: Our Place in the World

[1] Murie, Olaus J. *A Field Guide to Animal Tracks,* Second Edition. Boston: Houghton Mifflin, 1975, 2.

[2] This capable fungus is discussed in the fascinating book by Mark Plotkin, *Medicine Quest: In Search of Nature's Healing Secrets,* New York: Viking, 2000. Plotkin admires *Cordyceps* not only for its original and highly complex reproductive strategy, but also because, as an afterthought, it makes important pharmaceuticals that benefit people, as the Chinese discovered long ago. If it didn't benefit people, we might very well know nothing about it, as all too many living things must interface with people before we are inspired to investigate them.

[3] Sperduto, David D., William F. Nichols, Katherine F. Crowley, and Douglas A Bechted. "Black Gum (*Nyssa sylvatica* Marsh) in New Hampshire." Report submitted to the U. S. Environmental Protection Agency by the New Hampshire Natural Heritage Inventory, DRED Division of Forests & Lands (Concord, NH) and the Nature Conservancy, April 2000, 6.

Acknowledgments

I am indebted to many people for this book. Our wonderful neighbor, Bob Metcalfe, took the photo of a doe licking the ear of her daughter, and let us use it for the cover. Sy Montgomery has reviewed the manuscript time and again with encouragement and splendid suggestions. I am very grateful for her help and for her friendship. I am also very grateful to Howard Nelson for the use of "Camping Alone," one of his many splendid poems. For valuable information included in the book, I would like to thank Anne McBride, Ilisa Barbash, Hunt Dowse, Castle McLaughlin, Don Schrock, and Jasper Thomas. If there are mistakes in the book they are mine, needless to say, and certainly not theirs.

I am also grateful to my editor, Bruce Nichols, one of the best I've ever worked with, and to the copy editor, Shelly Perron, also one of the best I've ever worked with. And finally, as always, I am grateful to my agent, Ike Williams, and Hope Dennekamp.

Index